BUILD
UNIVERSES

James L. Kearns

Demonic Aliens

© 2022 **Europe Books**| London
www.europebooks.co.uk | info@europebooks.co.uk

ISBN 979-12-201-1962-7

First edition: March 2022
Distribution for the United Kingdom: **Vine House Distribution ltd**

Printed for Italy by *Rotomail Italia S.p.A. - Vignate (MI)*
Stampato presso *Rotomail Italia S.p.A. - Vignate (MI)*

DEMONIC ALIENS
JAMES L. KEARNS

TABLE OF CONTENTS

Chapter I
ALIENS ARE NOT "ALIENS!"

"And no marvel; for Satan himself is transformed into an angel of light." 2nd Corinthians 11:14

Thousands or even millions of people in the United States claim to have seen UFOs or UAPs (Unidentified Flying Objects or Unidentified Aerial Phenomenon) as they are now being called by government agencies. To many, they are aliens from some faraway planet to come to Earth to "help" us. I submit they are not what people think they are but are the same old same old demons from the very beginning of creation. Demons are fallen angels that were once in Heaven with God but rebelled and now inhabit the Earth in many different places. The fact that they travel in what seem to be "flying saucers" is just their way of getting around with their demonic "human-oid" bodies they are now inhabiting. Their lights are displayed over many parts of America and the world nearly every night now, as their time is short. Their plan is the same as it was from the very beginning, to take over the world and create demonic humanoid hybrids and get rid of us.

> There were giants in the earth in those days; and also after that, when the sons of God came in unto the daughters of men, and **they bare children to them, the same became mighty men which were of old, men of renown.** Genesis 6:4

Notice that the demonic entities mated with human women and created "alien-human" hybrids. They are doing the identical thing now with abductions and other practices they perform to create such beings. They are gathering genetic materials to make the demonic beings.

The demons are liars that can convince us they are from somewhere else in our universe. The US Government may in fact be assisting them to accomplish their goal. The government thinks they come from another planet, and we are helping them to survive and trading their technology for our genetic materials from humans and animals to create the alien-human hybrids. Area 51, as it is so named is the place where this "research" is most likely taking place. Since the crash at Roswell, New Mexico, on July 7, 1947, it has been going on. As I write this book, the government is supposed to reveal what we know about UFOs, but they will leave this part out. They might even show the craft we have from the crashes, but never reveal the aliens we found piloting the ships. They will plead complete ignorance about where these vehicles come from and try to say they come from one of the foreign countries with high technology, but will never admit what they really are, craft piloted by demonic entities that were once fallen angels, but now are masquerading as aliens from the Planet X!

Do I really need to rehash all the stuff we know about the Roswell Crash and that we actually found at least one of the so-called aliens still alive? Or how about the fact that President Eisenhower actually met with them at one time? We probably even signed some type of treaty to trade off their technology for genetic materials so they can make more alien bodies. I have no quotes to show he did, but there are rumors. And what about the fact that President Kennedy was actually assassinated because he was going to reveal everything about the aliens? Or that Trump lost the last election because he was going to reveal about them, and the deep state could not have that information revealed. You can see where this is going, I believe the demons are in charge of our government, and they are calling all the shots. Where is any proof of this allegation? First of all, look again at what the Bible says:

> In whom **the god of this world** hath blinded the minds of them which believe not, lest the light of the glorious gospel of Christ, who is the image of God, should shine unto them. 2nd Corinthians 4:4

Who is the god of this world? Satan and his demons are actually in control of the world, and therefore they are in control of governments. But others have said that these alien crafts are not alien but demonic. World-renowned pastor and teacher Chuck Missler stated:

"I for one do know the truth, the grey aliens are demons, and as it says in John 8:32 – 'The truth will set you free".[1]

And again:

"But don't be fooled, these are not aliens as they want you to believe. These Grey aliens are demons…"[2]

So, you see, I am not alone in my assertion that these so-called aliens are something other than aliens from some faraway planet in our vast universe. How could we prove such a claim anyway? The most likely scenario is these alien creatures are from right here! The same demons that were always here at the beginning of creation, are still here now! They are doing the same things as

---

[1] From "Real Unexplained Mysteries" https://realunexplainedmysteries.com/grey-aliens-are-de-mons#:~:text=Chuck%20Missler%20known%20for%20his%20notable%20work%20on,signs%20of%20the%20soon%20approaching%20end%20of%20times. Accessed 6.22. 2021

[2] From "Real Unexplained Mysteries" https://realunexplainedmysteries.com/grey-aliens-are-de-mons#:~:text=Chuck%20Missler%20known%20for%20his%20notable%20work%20on,signs%20of%20the%20soon%20approaching%20end%20of%20times. Accessed 6.22. 2021

before, creating creatures that will undermine the perfect creation of God, and corrupt man into something we cannot even imagine. Other pastors and teachers have voiced this same sentiment, that these aliens are in fact demons. From the beginning, their goal was to usurp God from being the creator of the world and convince man that we came into being by some random chance of evolution, or even by some alien race "seeding" the world. So-called scientists come up with all kinds of weird theories to explain our existence, rather than to believe the only viable one, that there has to have been a creator, and not some chance happening to create us and other alien forms of life. If it is so easy, why not on the Moon, on Mars, with abundant life forms? Not so fast! It didn't happen in these places, because it didn't happen anywhere at all! If we find life anywhere else at all, it is the result of creation by God and not the product of evolution. The angels were a creation by God, and we were also created by God. Aliens, if they exist, must have been created by God. The Bible doesn't talk about them. It talks about demons and angels, so that is an indication that no aliens exist, but demons and angels do. Many other pastors, not just obscure right-wing wackos, will tell us the same thing.

Angels take on physical bodies from time to time and have always had that ability, so far as we know. St. Thomas Aquinas states it is an emphatic yes that angels do exist:

"The term *extraterrestrial intelligence* is almost exactly what St. Thomas Aquinas would have used for angels. He explains that angels are nonmaterial intelligent beings. They have no corporeal bodies and are pure intelligence. However, Aquinas says that angels can assume physical bodies. They do not take over an existing physical body but they "manipulate matter so as to assume a physical appearance that is visible yet consistent with angelic character" (*Summa Theologica*, q. 51, art. 2).

Angels, Aquinas says, take physical form in order to communicate God's message. They are also the main agents of God's action in the world. So, he says, "Although creativity cannot belong to them [since only God can create from nothing] angels are nevertheless the chief ministers employed by God in the governance of the universe, in securing His own glory and in distributing His goodness to all creation." In other words, angels are God's secret agents in the world—sent to do his bidding, communicate with us and watch over us.

In a telling warning, Aquinas also reminds us that not all angels are good, and that remarkable phenomena may be produced by the action of bad angels."[3]

So, since demons can inhabit material and physical bodies, the "aliens" are not aliens, but just demons trying to deceive us they are from some faraway planet, and they are here to "help" us. *The Twilight Zone* episode I saw years ago was an indication of what this "help" would be like. In the television show, the aliens arrive and give us a book from which our greatest minds were only able to decipher the first letters of the title which was "To Serve Man." The people thought it was great that these aliens were here to serve us. It wasn't until they deciphered more of the book, that they discovered it was a **cookbook!**

When the real aliens arrive, and they are revealed, the world will go crazy thinking these aliens are here to help

---

[3] From "Angels and Aliens," by FR. Dwight Longenecker, https://www.catholic.com/magazine/print-edition/angels-and-aliens accessed 6.23.2021.

us, they are our "saviors." In reality, they are here to destroy man and corrupt the creation by God into something else. Look back at the scripture there from Genesis 6:4, "…**the sons of God came in unto the daughters of men, and they bare children to them, the same became mighty men which were of old, men of renown.**" So how could they do this unless they were able to inhabit physical bodies? Today, they are using a different method, gathering genetic material and growing alien-humanoid bodies, possibly with the assistance of the US Government at Area 51. They are trading their technology for the genetic materials needed to grow the bodies.

Black helicopters accompany the cattle mutilations that are happening on a massive scale, where specific parts of cattle are gathered, cutting genetic material out surgically to gather material for this purpose. Not many people would think aliens are traveling in helicopters, but "Men in Black" would. Why would the US Government somehow be involved in such a practice? Why not just buy the cattle from the ranchers and then do whatever experiments they need to do? The aliens dictate how the mutilations are going to proceed. *The History Channel* reports all about these mutilations and they are rampant:

"The bovine corpses stunned the ranchers who found them. The animals' ears, eyes, udders, anuses, sex organs and tongues had routinely been removed, seemingly with a sharp, clean instrument. Their carcasses had been drained of blood. No tracks or footprints were found in the immediate vicinity—nor were any of the usual opportunistic scavengers.

Between April and October of 1975, nearly 200 cases of cattle mutilation were reported in the state of Colorado alone. Far from being mere tabloid fodder, it had become a nationally recognized issue: That year, the Colorado Associated Press voted it the state's number one story. Colorado's then-senator Floyd Haskell asked the Federal Bureau of Investigation to get involved."[4]

This is not just happening here in the United States, but in other countries as well. The prevailing determination is it is known predators or religious satanic cults that are the cause of these anomalies. If it ended with a few cattle being killed, that would be one thing, but it does not end there. These demonic aliens are abducting humans and collecting genetic material and producing alien-humanoid hybrids. There are literally thousands of cases where men and women, even children have been abducted by these alien beings, usually the Gray's, but at other times other types of aliens. Betty and Barny Hill, Travis Walton, and thousands of recorded instances when people claim to have been abducted. Harvard professor John Mack states it emphatically: [5]

---

[4] *The History Channel* "The Mysterious History of Cattle Mutilation," https://www.history.com/news/cattle-mutilation-1970s-skin-walker-ranch-ufos Accessed 6.23.2021.

[5] "Harvard Psychiatrist On Alien Abductions/Contact: "Yes, It's Both Literally & Physically Happening." https://www.collective-evolution.com/2015/06/28/harvard-psychiatrist-on-alien-abductionscontact-yes-its-both-literally-physically-happening/ Accessed 6.23.2021.

"Yes, it's both. It's both literally, physically happening to a degree; and it's also some kind of psychological, spiritual experience occurring and originating perhaps in another dimension. And so, the phenomenon stretches us, or it asks us to stretch to open to realities that are not simply the literal physical world, but to extend to the possibility that there are other unseen realities from which our consciousness, our, if you will, learning processes over the past several hundred years have closed us off."

So, you are still a skeptic. You are not convinced of such a diabolical scheme by aliens, demons, fairies, or anyone/anything else. But hold on and you will find out it is soon going to be revealed that these aliens are here now. As I write this book, it is this month (June 2021) when the US Government is supposed to come clean about the aliens, but they will only reveal part of the information, not the abduction parts, nor the cattle mutilation, nor the other evil that is happening now at the hands of these demons. They will and have said that these UFOs and UAPs are real, but apart from that will not speculate or give full disclosure about them. They will conclude we just do not know what they are, but that they do not seem to be harmful, and they must be benevolent. I contend they are not benevolent, and we do know what they are: demonic entities, not aliens from some other planet, which are here to take over the world and destroy man as we now know ourselves. We will be changed into them if we do not take the steps to fight them. Other scientists and pastors have known about demonic possession and other occurrences that will make your head shake with denial, but it is true!

Even now you are being controlled by these Demonic Aliens in your daily life! You think you have everything under control, but what about the wrong things you do? Everyone is a sinner, and we continuously sin every day in ways we do not even think about. If you are a student, you cheat at least a little on your tests, you do not do your homework as you should, you disobey your parents constantly. If you are employed, you stretch the truth to your clients or you cut corners for your boss, and you take breaks when you should not and cheat your employer. You stretched the truth about yourself to get the job you have now, and these are only part of the wrong things we all do every day. We yell at our wives or husbands, spank the kids for no reason, kick the dog for barking too much, and the list goes on and on! This does not even touch your problems you may have with speeding on the highway, alcohol, drugs or pornography, or something even worse. Where does all this come from? I used to say it is from Satan and demons, but now I am coming to realize it is from the Demonic Aliens, which are the real demons.

Thousands of cases of abductions have found that people even have implants somewhere on their bodies. What in the world are they? When analyzed they are found to be many times constructed of materials, not from the Earth! The doctors who find them have no idea what they are for, but they are often emitting a signal. Obviously, they are some forms of tracking device! [6]

----

[6] From "An American Surgeon and Expert on Alien Implants," by Brent Swancer. https://mysteriousuniverse.org/2020/10/an-american-surgeon-and-expert-on-alien-implants/ Accessed 6.23.2021

Roger Krevin Leir started out as a well-respected podiatric surgeon from California, who was not always known for his involvement with aliens and UFOs. He gradually picked up an interest in Ufology over the years and became a member of his local chapter of the Mutual UFO Network (MUFON), attending conferences, but at the time it was more a hobby than anything else. It was at one conference in particular that his life would be changed forever. In 1995, Leir attended a MUFON conference and was presented the X-rays of a woman who claimed to have been abducted by aliens. The X-rays were strange, and since Leir was a doctor, they were shown to him to see what he thought. They appeared to show some sort of small metal objects embedded within her, and he went about having them surgically removed out of sheer curiosity. What he found was something very strange going on. He decided to devote his time to studying this bizarre phenomenon, and would open a practice especially for this purpose, as well as a nonprofit What he found was that the objects were two tiny metal shards that when analyzed by a laboratory showed them to be composed of aluminum and a type of iron typically found in meteorites. There could be found no traces of scars to show a point of insertion, nor was there any inflammatory tissue or rejection reaction by the body to these foreign objects. How did these objects get there and what were they? Leir was perplexed, and when he removed similar objects from a second patient, he knew that there was research organization dedicated to it, called A&S Research, even hiring a dentist, a radiologist and a general surgeon to be a part of his team. What he would find over the course of at least 17 such implant extractions over his career would open his eyes to a whole new, bizarre world.

Now, you may not have been implanted yet, but the day is coming when everyone living will, according to the Bible, be unable to buy or sell without the "Mark of the Beast." Obviously, this is the precursor of this process! Take the "implant" or die!

16 And he causeth all, both small and great, rich and poor, free and bond, to receive a mark in their right hand, or in their foreheads:
17 And that no man might buy or sell, save he that had the mark, or the name of the beast, or the number of his name.
Revelation 13: 16-17

So, you can see how this is in fact a correlation to the upcoming time when, after the Demonic Aliens are revealed to the world as creatures from another planet and how they are here to help us. All you need to do is take the implant, you will be healed of your diseases, and you won't die! If you don't take it, you will starve to death! Now all this is a moot point if Jesus returns and takes you out of this before it happens. You have to be a Christian before the Demonic Aliens return, or you will go through this process and be condemned to Hell if you take the "Mark of the Beast!"

What if the Bible is true about the fact that we will all have to take the mark, or we won't be able to buy or sell, so you can't eat, because you can't buy food? And who is controlling these "marks?" Why the Demonic Aliens, of course! Don't think for one minute you can just hide from them in a bunker or cellar beneath your home, these creatures will find you! No ship on the ocean, no island in the Pacific, no bunker or dense forest, will be safe! They will find you, abduct you, and force you to take the mark or be beheaded. The time is very short before all these things I am writing about here will come to pass. The US Government is again preparing this month to say that UFOs or what they call UAPs are in fact real! They won't yet include all the other stuff, but soon the information will leak out, the alien mother ship will land at the

White House, and they will step out for all the world to see! When this happens there will be no escape! My view is just after the Rapture of the Church, it will happen. The Demonic Aliens will say they have taken all those people to their planet, and they are here to help us also! Never mind that it was the Rapture, the world will believe the lie! If you are still here, the Tribulation has begun!

The year 2020 was the Covid-19 Pandemic that was a precursor of what is to come later on. Everyone had to comply with the lockdowns, wear masks or be arrested, travel only when necessary, get the vaccination shot in order to travel on a plane, practice social distancing, and the list goes on and on. They were even going to make you get a passport that shows you were vaccinated, in order to travel at all. I did not get the vaccination. I did not trust the government. I felt there may be some kind of a "tracker" in the shot to show who got it and who didn't. It would really make it easy for the Demonic Aliens to know who they can control and who they couldn't. If you got the shot, relax, it was not the "Mark of the Beast," but it is a precursor of what is to come. Soon, the same type of occurrence will happen again, and this time they will force you to get the "shot," except this time it will be the dreaded "Mark of the Beast!"

The US Government has FEMA camps set up all over the country. The barbed wire fence top is for both sides getting in or getting out. What do they need these camps for? Obviously for rounding up anyone who won't comply with the orders to take the shot, which in this new case, it will actually be the "Mark of the Beast." Not only is this true, but the US Military has already purchased 30,000 guillotines ready for just the purpose of chopping off heads! There is no other purpose for such a device! There is a House Bill to verify the purchase!

Georgia House of Representatives - 1995/1996 Sessions

HB 1274 - Death penalty; guillotine provisions

Code Sections - 17-10-38/ 17-10-44 A BILL TO BE ENTITLED AN ACT

1- 1 To amend Article 2 of Chapter 10 of Title 17 of the Official

1- 2 Code of Georgia Annotated, relating to the **death penalty**

1- 3 generally, so as to provide a statement of legislative

1- 4 policy; to provide for **death by guillotine**; to provide for

1- 5 applicability; to repeal conflicting laws; and for other

1- 6 purposes.

SECTION 1.

1- 8 The General Assembly finds that while prisoners condemned to

1- 9 death may wish to donate one or more of their **organs for**

1-10 **transplant**, any such desire is thwarted by the fact that

1-11 **electrocution makes all such organs unsuitable for**

1-12 **transplant**. The intent of the General Assembly in enacting

1-13 this legislation is to provide for a method of execution

1-14 which is compatible with the **donation of organs** by a 1-15 condemned prisoner. [7]

At the final portion of the bill, it says it is to be used on a condemned prisoner. That would be **YOU** if you refuse to take the mark.

Further, the report states that:

"The information we received is that 15,000 are currently stored in Georgia and 15,000 in Montana."[8]

There are nowhere near 30,000 prisons in the United States. Why in the world would we need 30,000? Obviously, there are going to be a lot of beheadings going on during the time when they try to force people to accept the mark.

Everything is falling right into place. We are rapidly and inescapably moving to a time when it will be just like the Bible says it will be. Who would have thought a few years ago that it would go down this far this quickly? I have a contention that those that are in power are controlled more by these Demonic Aliens than the general population. How would you know if it were true or not? They don't even know what they are doing sometimes. The fact that politicians could be in favor of allowing babies to be killed, right up until and even after birth, is an indication that they are not thinking with a clear mind. There has to be something wrong with them! Demonic Aliens have them under some *Jedi mind trick!* It also

---

[7] From "Why Did the U.S. Government Purchase 30,000 Guillotines?" By Michelle H. Published 4.29.2017.
https://www.linkedin.com/pulse/why-did-us-government-purchase-30000-guillotines-melinda-h/ Accessed 6.24.2021
[8] From "Why Did the U.S. Government Purchase 30,000 Guillotines?" By Michelle H. Published April 29, 2017.
https://www.linkedin.com/pulse/why-did-us-government-purchase-30000-guillotines-melinda-h/ Accessed 6.24.2021

seems that the richer a person is, the more he is controlled by evil forces. Either that or God. There are only two different ways a person can be controlled, by God or the devil. The Demonic Aliens are manipulating and controlling them without their knowledge. How would we know if they aren't? They don't have to be rich either to be controlled by demonic forces.

Ted Bundy was a charismatic person who could have been a successful lawyer. Some people even thought he could have been the next Kennedy-type political figure. He was controlled by evil instead to the point that he raped and murdered many women. He credited pornography as the reason he got into this practice, but it is obvious that he was controlled by demons, not God! Denying this fact is where we refuse to accept what is happing now, the demons (Demonic Aliens) take control of the people they can, even without their knowledge or consent, and compel them to do all manner of evil acts, and we overlook the obvious reason for their actions. They are not just somehow evil but are evil because of demonic control. Today we are all too sophisticated to believe in demons or God. In the movie, *The Usual Suspects* the main character (Kevin Spacy as Verbal Kint and later revealed to be Keiser Sosa) says "The greatest thing the devil ever did, was to convince the world he doesn't exist." [9]

When these Demonic Aliens land at the White House, they will not say they are demons, but they will surely lie and say they are from say, the Planet Krypton (anything but the truth), out in some other solar system, far enough away that we can't possibly verify it. The world will

---

[9] *The Usual Suspects*, Released by Polygram Film Entertainment, 1995.

blindly believe it all. It will stop dead in the water any belief in God or demons, and send the masses spiraling out of control down a track to worship them and not God. They will heal everyone and tell us we all have to get the implant they offer, and then we will be healed. It will change us; we will be more easily controlled and be able to be tracked. If we don't get the implant, we will be considered an enemy and then we will be arrested, taken to FEMA camps and if we absolutely refuse the implant, well, that is what the guillotines are for! There are also stored at locations in Georgia and other states, plastic caskets in the hundreds of thousands, large enough to hold up to three grown adults in each one! Now they will try to debunk this idea, and say they are vaults and not coffins, but why in the world do we need 500,000 of them? Check it out - they are stored in Madison, GA.

Now before I became a Christian, I had lots of issues as a child. I was very shy and scared of girls. I think a lot of young boys have that problem, but I was very afraid. This led to issues with pornography, which I am not proud of, but it needs to be told. I didn't even start dating until I became a Christian, on January 20, 1980. I was 25 years old. I will let you do the math. I bring this up because I had an experience where I was attempted to be controlled by a demonic force after I became a Christian. I wrote about it in my first book. *Finding Proof of Jesus.* Here is the account from the book:

Now, what I am about to say is supernatural. I was living in Florida, in a small, one-bedroom apartment. I am ashamed to say that at this time in my life and as a young Christian, I was involved in pornography. This involvement is reason for what happened next. Pornography is very dangerous and will destroy you if you do not repent. On three separate nights, I woke up and felt a very heavy weight on top of me. I could not move at all, but I could hear someone breathing. I was the only one in the room, as I lived alone.

It subsided each time after a few moments, but it was very scary. I had been saved for a little while at that time, but now I realize what that thing was—a demon, trying to control me and take over my body. After the third time, a miracle happened. I rose from sleep and looked toward the mirror above the dresser. From the angle of my bed, I could see an image in the mirror of Jesus Christ, hanging on the cross. I want to make clear that I did not see God but only a reflection in a mirror. As the Bible says in John 1:18, "No man has seen God at any time, the only begotten Son, which is in the Bosom of the Father, he hath declared him." Then, that thing left me and has never returned.[10]

Now, the only reason that I am not walking around as someone else, as a demonic entity, is because I had recently become a Christian, otherwise, I would be "captured" by these Demonic Aliens, and unaware of being controlled, but thinking I am in control, but in reality, totally possessed and controlled by Demonic Aliens. You may already be one of "them," controlled by the Demonic Alien Overlords, and you don't even know it! If you are not a Christian, you are in effect, just that. If you don't serve God, you are serving the demons, which I contend are the Demonic Aliens which will soon be revealed!

---

[10] *Finding Proof of Jesus*, James L. Kearns, Published by WestBow, 2016 p. 9, 10.

Chapter 2
RUSSIAN EXPERIMENTS

You might not be aware of all the atrocities committed during WWII, but there were experiments to try to "bread" an ape with a man. Apes are much stronger than men. If you have an obedient soldier that was 1.5 times more powerful than humans, you have a significant advantage over your enemy. The experiments failed but after the war, it was discovered that these types of hybrid experiments were actually funded by Stalin. The scientific community will tell you that it was not for the purpose of trying to create such a creature for war, but what is the real purpose? Ethically it is not warranted to produce such a hybrid creature. The United States Military is also involved, I guarantee it, with a similar type of research, because if we don't do it, another country will, and we would be left holding the bag with inferior soldiers if there was ever a war. The comic book character Captain America is not far-fetched at all.

"…the claim comes from a 2002 paper in the academic journal *Science in Context,* by the Russian historian of science Kirill Rossiianov. Rossiianov's study follows the ill-fated attempt by the Russian physiologist Il'ya Ivanov to crossbreed humans with anthropoid apes."[11]

To be fair, they say it was not for this purpose, but I will ask, what other purpose is there? And now, the United States is involved, up to their lying teeth, in this same type of research, only this time it is with aliens!

---

[11] "Scientific America," "Scientific Ethics and Stalin's Ape-man Warriors," From https://blogs.scientificamerican.com/primate-diaries/stalins-ape-man-superwarriors/ Accessed 6.26.2021

SPACE ALIENS BREEDING WITH HUMANS

There are several people who believe that we are being visited by space aliens and they are "breeding with humans!" From the article stating just that, "Space aliens are breeding with humans, university instructor says. Scientists say otherwise," we have mainline people believing it is happening.[12]

> …an instructor at the University of Oxford in England believes the abductions are real. Young-hae Chi, who teaches Korean at the university, also claims to know what the aliens have in mind. In lectures given at the university, he says they're <u>creating alien-human hybrids as a hedge against climate change</u>. To support his unorthodox theory, Chi notes that for several decades the number of reported alien abductions has risen. He bases this statement on the work of David Jacobs, a retired Temple University historian who has published several books on ufology and who runs the International Center for Abduction Research.

The black helicopters that accompany the cattle mutilation occurrences are an indication that the US Government is absolutely involved in this process. As I write this evening, the government has stated that we still just don't know what the UFOs or UAPs are, like I said, they will not admit that we know exactly what they are and why they are here. They also stated there is no indication they are from another planet. They continue to say we don't know what they are.

However, while there's no evidence the objects are extraterrestrial, senior government officials who briefed reporters said Friday that nearly all of the incidents

---

[12] "Space aliens are breeding with humans, university instructor says. Scientists say otherwise," by Seth Shostak, May 26, 2019. https://www.nbcnews.com/mach/science/space-aliens-are-breeding-humans-university-instructor-says-scientists-say-ncna1008971 Accessed 6.26.2021.

investigated remain unexplained, citing the significant limitations of available data and reporting.

"Of the 144 reports that we are dealing with here, we have no clear indications that there's any non-terrestrial explanation for them," a senior government official said, nixing the possibility of exotic sightings. "But again, we will go wherever the data takes us on this."[13]

They are continuing to cover everything up and will refuse to release any information that leads the public down the path as to what they actually are – craft piloted and controlled by Demonic Alien beings, not from another planet, but right here on Earth! To their credit, they did say there was no indication they were extraterrestrial. In this statement they are correct.

ADMIRAL BYRD AND THE DISCOVERY OF AN ALIEN BASE IN ANTARCTICA

These creatures are from right here in secret bases on the Earth. Admiral Byrd, on the expedition to Antarctica, found one of the bases and met with the inhabitants at the base.

"Interestingly, my own introduction to allegations of Byrd meeting strange beings while visiting the Antarctic came from an odd story shared with me by a friend, rather than any alleged secret diaries or other fabled texts. According to the story my friend told, a family member of his who had known a member of Byrd's expedition had told stories **of "blue-skinned people from underground"** that were encountered by Byrd and his

---

[13] From "Pentagon task force's UFO report released – many cases remain unexplained," by Nicole Sganga, David Martin, Olivia Gazis and Eleanor Watson. https://www.cbsnews.com/news/pentagon-ufo-report-released-many-uap-cases-remain-unexplained/ June 25, 2021. Accessed 6.26.2021

company. This odd tale remains somewhat in keeping with the aforementioned account of an underground race of beings led by an esoteric "Master"; on the other hand, it differs greatly from the sort of theories held by some UFO buffs, who allege that none other than escaped Nazis had taken refuge at the Southern Pole, along with their curious saucer aircraft."[14]

There are many different types of Demonic Aliens, but the Grays seem to be the most common. There may be as many as 36[15] or more different types of alien creatures, according to some Internet searches, but again, they are not aliens, but demons. They have alien bases under the oceans, the Artic, and at other locations on the Earth. The whole idea that these creatures come from outer space is not likely, because the US Navy is now finding these craft travel under the water, as well as in the air. They obviously have bases on areas on land and sea. The recently released classified chase of an "alien" craft ended when it submerged under the ocean. Other people have seen many times when these crafts went under bodies of water. So, we can assume the bases are right under our noses and not "out there" somewhere. Remember, the released report said there is no indication they are from another planet.

---

[14] From "The Truth About Admiral Byrd's 'Bitter Reality' at Earth's End," 2.10.2014, by Micah Hanks. https://www.gralienreport.com/ufos/the-odd-exploits-of-admiral-byrd-bitter-reality-at-earths-end/ Accessed 6.26.2021.
[15] From "Could there be 36 species of aliens out there?" Newsround, 6.18.2020. https://www.bbc.co.uk/newsround/53068978 Accessed 6.28.2021.

They are here on the Earth. The Bible tells us that. They are not from some planet billions of miles away. Consider this Scripture:

> And the LORD said unto Satan, Whence comest thou? Then Satan answered the LORD, and said, **From going to and fro in the earth, and from walking up and down in it.** Job 1:7

So, we know that the devil is right here, in bases all over the Earth, hidden in plain sight. They are usually under the Earth or sea. One such area is Dulce, New Mexico. The rumor is this, secret underground bases exist on the Earth. One article from "Discovery," "Allegedly, There Is a Secret Underground Alien Base in Dulce, New Mexico," accounts:

On the surface, Dulce, New Mexico is just a small southwestern town. It doesn't even have a traffic light. But according to the most bizarre rumors, this little town is just a cap on a gargantuan underground facility that is home to unimaginable experiments and technologies. For the tinfoil hat-wearers, there's a whole world underneath Dulce — a secret, high-tech one filled with aliens.[16]

We are allegedly involved in the development of human-alien hybrids. The article continues:

"According to conspiracy theorists, the Dulce subterranean base is a seven-story compound beneath Dulce, New Mexico that houses human-animal hybrids, human-alien hybrids, and extremely advanced technologies. They say even been the site of alien

---

[16] From "Discovery," "Allegedly, There Is a Secret Underground Alien Base in Dulce, New Mexico," By Joanie Faletto. https://www.discovery.com/exploration/Secret-Underground-Alien-Base-Dulce-New-Mexico 8.1.2019, Accessed 7.5.2021.

wars. You know, the usual. It hasn't been called the Roswell of Northern New Mexico for nothing."[17]

Area 51 is of course another area where if you try to enter that area, deadly force is authorized, and you will be killed. No one I know of has really tried to break that security, and if they have, they have never been heard of again. So, we have the Artic, attested to by Admiral Byrd, Dulce, New Mexico, Area 51, and also the underwater bases. The ones underwater are more difficult to get to, obviously, but they are there as well.

SECRET UNDERWATER BASES FOR DEMONIC ALIENS

There is ample evidence of underwater bases for these Demonic Aliens. One of many published articles states:

In August 1981, five witnesses who were driving alongside Lake Ontario early one evening saw a dome-shaped craft flying over the water. They followed the craft for some time before seeing it begin to descend and enter the water, disappearing from their sight. Perhaps the stories of an underwater alien base, stem from the 1977 book *The Great Lakes Triangle* by Jay Gourley, who made note that many planes and people had disappeared over Lake Ontario, not to mention the many UFO sightings of the area.

Another book, *Underground Alien Bases*, released in 2012 and written by the somewhat strange "Commander X," has also perpetuated the legend of the alien base under the lake. It features within its pages several accounts of sightings on and around the lake and the assertion that

---

[17] From "Discovery," "Allegedly, There Is a Secret Underground Alien Base in Dulce, New Mexico," By Joanie Faletto. https://www.discovery.com/exploration/Secret-Underground-Alien-Base-Dulce-New-Mexico 8.1.2019, Accessed 7.5.2021.

an alien fortress of one kind or another lies under the water. However, none of the accounts can be verified by a secondary source and so are left open to debate as to how reliable they are.[18]

According to the article, there may be as many as ten known alien bases. It is unknown how many of these bases actually are operated with the knowledge of the US Government.

THEY ARE HERE TO "HELP" US!

The idea that aliens from another planet traveled billions of miles and thousands of light-years to this planet to help little old me is just absurd. They abduct people by night, take samples of eggs and seaman, conduct all manner of weird medical tests on us, implant devices in different areas on our bodies, and who knows what other evil procedures they do! They only kill a "few" of our cattle, terrorize us by the sight of their crafts flying in the sky, and do all this under the cover of darkness and clandestinely. That is not helping! Again, the US Government is obviously involved in this whole process, with the cover-ups and the black helicopters hovering over the sight where these crafts are seen. They say, "We are not going to hurt you! Relax and let us take some blood and implant a tracker in you and get a sample of your sperm or eggs!" Not going to hurt us, indeed!

PROOF OF THE UNITED STATES BEING INVOLVED IN CATTLE MUTILATIONS

---

[18] From "Stillness in the Storm. An Agent of Conscious Evolution," "10 Alleged Alien Underground Bases" https://stillness-inthestorm.com/2020/02/10-alleged-underwater-alien-ufo-bases/ Accessed 7.5.2021

Countless sightings of these alien crafts are accompanied by the black helicopters. It has been happening for decades. From the article by Nick Redfern, "The Mystery of the 'Black Helicopters': UFOs and Cattle Mutilation," he states emphatically what is going on.

Sightings of what have become known as "Black Helicopters" and "Phantom Helicopters" have been reported for decades. There's no doubt, however, that the 1970s was the period when things really took off. In the latter part of 1973, a series of distinctly unusual events began to quietly unfold above the green fields and rolling hills of northern England that subsequently triggered a strange, and still-ongoing operation coordinated by an elite division of the British Police Force. Remarkably, that operation was (and, to an extent, still is) designed to carefully monitor the activities of certain key players within the public UFO research arena in Britain. And it all began with something known as the phantom helicopter. Of the varied elements that make up what is popularly known as the "UFO Phenomenon," there can be few so strange as the phantom helicopter and its close relative, the even-more-sinister black helicopter. For at least four decades, numerous people throughout the world have reported seeing helicopters, very often completely black in color and with no identifying markings, in areas that have been subjected to intense UFO activity. This has led a number of commentators to speculate that the helicopters are operated by covert groups from within the military-industrial complex and that they are involved in a surreptitious UFO monitoring program.[19]

These "black helicopters" are seen in the vicinity of these sightings and cattle mutilations as almost standard procedure. Aliens don't travel from other worlds in black helicopters, so we can trust the military is involved with the mutilations. There are so many UFO sightings that have the sighting of the black helicopters and the "Men in Black", that I don't need to put any more quotations in this book. The movie would be funny, if it was not real, at least about the part of the Men in Black. They are men in dark black suits that try to hide what the people saw, and from them coming forward about the sightings, by intimidation, fear-mongering people into submission, and other tactics to cover up the issue for these Demonic Aliens. The demons are in control of what is happening in our world today, and we are their "guinea pigs"!

THE ROSWELL CRASH AND THE COVER UP

I don't need to rehash all the details about The Roswell Crash, but to be fair, if you never heard about the particulars of what happened, and how it was later covered up, there is a good movie out there called "Roswell," that I think is pretty close to what happened. The crash happened on the ranch of J. B. Foster where Max

---

[19] "The Mystery of the 'Black Helicopters': UFOs and Cattle Mutilation," Nick Redfern, March 20, 2021. https://mysteriousuniverse.org/2021/03/the-mystery-of-the-black-helicopters-ufos-and-cattle-mutilations/ Accessed 6.28.2021.

Brazel worked in early July of 1947.[20] The local paper in Roswell, New Mexico reported a flying saucer had crashed there at first after the military reported it as such. It was then quickly covered up by the military and they made the paper print a retraction that it was just a weather balloon. But the huge problem with this was the military spent days at the crash site gathering all the materials that had crashed. They then sent the materials off to a base in Dayton, Ohio. If it were just a weather balloon, why all the activity at the site, and then why did the military come by later to people's houses and threaten witnesses, even with imprisonment (some say even death) if they spoke up, and demand they keep quiet about what they saw. The rumor mill started back then, and even that there were bodies of alien creatures! One of them even survived and lived for years after the crash. The government of course covered all this up and made-up stories later about how it was a classified type of balloon that crashed, Project Mogul,[21] and it had dummies attached to it, which was to quash the rumor of the alien bodies. How many people would believe a dummy was an alien body? They think we are really stupid!

---

[20] Space.com, "Roswell UFO crash: What is the truth behind the 'flying saucer' incident? By David Crooks, 5.5.2021. https://www.msn.com/en-us/news/world/roswell-ufo-crash-what-is-the-truth-behind-the-flying-saucer-incident/ar-BB1gog3T Accessed 6.28.2021.

[21] Space.com, "Roswell UFO crash: What is the truth behind the 'flying saucer' incident? By David Crooks, 5.5.2021. https://www.msn.com/en-us/news/world/roswell-ufo-crash-what-is-the-truth-behind-the-flying-saucer-incident/ar-BB1gog3T Accessed 6.28. 2021.

All the hubbub about aliens is a moot point, though. They are not aliens, but demonic entities that have obtained humanoid bodies, for the purpose of taking over God's creation. Before they mated sexually with humans to produce hybrid humans like the scripture says in Genesis 6, but now they are possibly both doing that, by injecting alien embryos into women, if not also sexually molesting them, and genetically manipulating our species! There are cases where human women give birth to tremendously grotesque babies after they report they were abducted. But many times, the babies survive and are just different. In the article "Alien Hybrids and Humanoids," it shows many gross pictures of the hybrids, and the accompanying video detail the purpose of the aliens as making us "better."

The hybrids are not really aliens, but a mixed breed between the Grays and humans. They came about when the Grays started doing gene DNA manipulation many years ago starting in at least biblical times supposedly to replenish their own thinning gene pool after centuries of over cloning.

Another theory as to why the Grays started cloning is more cynical, being that they started to place the alien hybrids into Earths society to help them overthrow the Government and take over from the inside rather than a full-fledged attack. Some even speculate that many of the alien hybrids on Earth don't know that they are of mixed nature and are like terrorist cells that can, and will be activated when the time is right for the master alien races (The Grays and the reptilians).[22]

---

[22] Alien UFO Research, "Alien Hybrids and Humanoids," https://alien-ufo-research.com/alien-hybrids/ Accessed 6.28.2021.

These so-called aliens are again, not aliens from another planet, but demons, fallen angels that went with Satan to try to take over the world back at the beginning. A third of the demons went with Satan according to the Bible. I am going to keep reiterating this fact throughout the book, until hopefully it finally sinks into your psyche that what we are dealing with in the whole UFO issue is Demonic Aliens, bent on destroying what God made and ruling it themselves. They want to usurp God, overthrow God, and Satan wants to be God. But, he has failed and will ultimately be destroyed and cast forever into Hell, but he will still keep trying, thinking that if he keeps at it, he will eventually succeed. The cosmic chess game for the control of the universe is still being played out, although a checkmate by God is inevitable. When Jesus died on the cross, that was the checkmate, but Satan still thinks he can win. He knows he will go to Hell, but he can still take as many of us with him when he goes. That is why he still plays the game, to take as many people with him as he can to Hell. So now, with the time getting closer to when the Rapture occurs, and the people left behind looking for an answer, the Demonic Aliens land at the White House and say, "We are here to help you! We have taken many people already to our home planet!" That will be the ultimate lie.

Proof of this is in the pudding, as they say. What would a benevolent super-advanced alien race have to gain from contacting us? Absolutely nothing, and to be sure, they wouldn't be here to help, but like the movie "Independence Day," they were here to kill us all! They just wanted us to die! In that movie when the heroes are entering the mother ship, they saw the aliens massing in troop strength, to go door to door to kill everyone that remained after they had defeated the world's military might. So here we have a word about gun control, we need them to battle the demons that will try to take over the world after the military is defeated. These creatures "die" in bodily form, just like we do, but their "souls" live forever. But regretfully, with the implant, we will have to cut off their heads to kill them. We can slow them down with an AK-47 or a good 9 mm pistol, but that is all it will do. So, where is any proof of what I just said? Look at the Bible again:

> [11] And I beheld another beast coming up out of the earth; and he had two horns like a lamb, and he spake as a dragon. [12] And he exerciseth all the power of the first beast before him, and causeth the earth and them which dwell therein to worship the first beast, whose deadly wound was healed. [13] And he doeth great wonders, so that he maketh fire come down from Heaven on the earth in the sight of men, [14] And deceiveth them that dwell on the earth by the means of those miracles which he had power to do in the sight of the beast; saying to them that dwell on the earth, that they should make an image to the beast, which had wound by a sword, and did live. [15] And he had power to give life unto the image of the beast, that the image of the beast should both speak, and cause that as many as would not worship the image of the beast should be killed. [16] And he causeth all, both small and great, rich and poor, free and bond, to receive a mark in their right hand, or in their foreheads: [17] And that no man might buy or sell, save he that had the mark, or the name of the beast, or the number of his name. [18] Here is wisdom. Let him that hath understanding count the number of the beast: for it is the number of a man; and his number is Six hundred threescore and six. Revelation 13: 11- 18.

Notice this creature doesn't die, but he is healed, as will be all the people that take the "Mark of the Beast". Then, they are changed into the demon-possessed alien creatures that will fight against God at the very end. Now, the main biblical eschatology is for all of this to happen after the Rapture, but what if the Rapture doesn't happen until these creatures are revealed and we have to fight them to keep from taking the mark? It is surely something to think about and to be prepared for, just in case. As bad as everything is getting now, either Jesus will come very soon now, or the world will turn into the Hell on Earth of the Tribulation. I believe that the return of Jesus in the Rapture is imminent, hence the writing of this book, to get as many people prepared for when that happens, but we must be ready for everything. If, and when the Demonic Aliens land on the White House lawn, it is the beginning of the end, even if Jesus has not returned yet.

Chapter 3
IT'S THE ZOMBIE APOCALYPSE!

So how do these Demonic Aliens cause us to "sin"? Who is to say if they are shooting some kind of "ray" down, targeted specifically for you that will make you do their bidding? There are radio and Internet waves all around us all the time and we don't even think about it. Cell phones use waves, and we talk to others through the air, so why not some similar type of wave or ray from these Demonic Aliens that tempt us to sin, or actually control us in other ways. Once they are truly revealed, you will have to get the implant, or be arrested and ultimately have your head cut off if you don't comply! Think I am kidding? We are already there! During the Covid-19 Pandemic, millions rushed to get the vaccine. If you didn't wear a mask and tried to enter a store or restaurant, you were arrested. I might be repeating myself here, but it bears repeating that we will be forced to do things we might not want to do. The implant to control us is already here, but even if what we have developed is not really the "Mark of the Beast," the Demonic Aliens will have their own. One article states:

Back in the year 2015 in Sweden, a high-tech office block started offering employees the option of getting a miniature sensor beneath their skin in place as a substitute for the swipe card to pay for food at the café or access amenities such printers. At the same time, a remote-control contraceptive chip was created by Massachusetts-based MicroCHIPS. This chip allowed the woman to turn shots of birth control medicine off and on. Also, digital RFID tattoos are obtainable for those who want to monitor their body by measuring temperature, UV exposure, hydration levels, and other bodily factors. In addition, the microchip technology is now even used in hospitals to diagnose and treat patients[23]

What most people don't realize or think about, is that soon, there will be a requirement for everyone to get this implant if they want to work, or if they want to buy food. Not as far-fetched as it sounds. Back when I first started working, the company decided that everyone had to have a checking account to get paid. You had to allow for direct deposit, or you wouldn't get a check. I went along with it blindly, what choice did I have? Here, too, most people won't give it a second thought about getting that implant. Lots of people already have. I did not get the vaccine this time because I had heard that it had a type of "tracker" in it so they could tell who got it and who did not. Even if that is not true, I should have the choice to decide if I want to get it or not. It caused the deaths of about 30 people a day that got it. I heard this on Fox News and also reported here in this article, that large numbers of people are dying from it. That is a large percentage of risk, where normally the rate is very low, but the percentage of people who die from this vaccine is not really reported.

"Between mid-December 2020, when the first COVID-19 shots were rolled out, and April 23, 2021, at which point between 95 million and 100 million Americans had received their COVID-19 shots, there were 3,544 reported deaths following COVID vaccination.

That's 182 more deaths than cited by Carlson. As of April 23, 2021, VAERS had also received 12,618 reports of serious adverse events. In total, 118,902 adverse event

---

[23] From "The Enterprise World", "Implanted Microchip: The Mark of the Beast?" https://www.theenterpriseworld.com/ implanted-microchip-mark-of-the-beast/ Accessed 6.29.2021

reports had been filed. If, like Carlson estimates, about 30 people per day are dying from the shots, these numbers will grow by the hundreds each week."[24]

The article continues:

"Folks, this is 70 times more deaths than the swine flu vaccine, which was halted. If this isn't insanity on steroids, please tell me what is. Maybe murder? This doesn't even include the deaths of thousands, and potentially tens of thousands of miscarriages, which is now becoming rapidly recognized as a possible complication of COVID-19 'vaccines.'"[25]

It is a large risk to get the vaccine. It is better to take steps that would keep me safe and hope for the best. Wear the mask when I am out in public, stay indoors when I can, wash my hands often, normal good hygiene procedures. To date. I didn't get the vaccine and I didn't get Covid-19, but if I do get it and die, that is better than getting a tracker implanted in my body. There are tweets that state that Bill Gates is launching implantable chips in the vaccine.[26] The powers that be will always deny such a claim, but how would we know anything about what is in it.

---

[24] "Undercurents," "How Many People Have Died From Covid-19 Vaccines?" https://undercurrents723949620.wordpress.com/2021/05/22/how-many-have-died-from-covid-vaccines/ Accessed 6.29.2021.

[25] Undercurents," "How Many People Have Died From Covid-19 Vaccines?" https://undercurrents723949620.wordpress.com/2021/05/22/how-many-have-died-from-covid-vaccines/ Accessed 6.29.2021.

[26] From "Trust Index: Does the COVID-19 vaccine contain a tracking device?" https://www.ksat.com/news/local/2021/01/12/trust-index-does-the-covid-19-vaccine-contain-a-tracking-device/ Accessed 6.29.2021.

When I went in the Navy, they put us in a line at boot camp and they shot us full of stuff. I had no idea what they were shooting me up with, but I had no choice but to take the shot as part of the military. If I refused, of course, I would have been taken to Captain's Mast and then court-martialed if I refused. I went along with the whole process like a sheep to the slaughter. You will too if it gets bad enough. You will have to eat, so you will get the implant, and then it will be all over for you! Then you will be changed into a creature that can't die! A zombie that will follow the orders of the Demonic Alien Overlords, and you won't even know you are being controlled. We are rapidly becoming a people that are controlled by forces outside of us. The riots that destroyed inner cities looted goods out of stores, destroyed court houses and police precincts, are an indication that people are not thinking about what they are doing, just doing it. But you say, "That is not me!" It will be you once you get the implant! You will be walking around like a normal person, thinking you are doing "your own thing," but in reality, you will be doing the bidding of these Demonic Aliens! They are implanting those implants in people, but also controlling us in other ways. The Bible talks about sin. I am powerless many times to resist doing what I am tempted to do. Often, I just do it without even thinking. I try to resist sometimes, but mostly I fail. You do too! How does the devil cause us to sin? Maybe it is from these Demonic Alien Overlords shooting the ray down to tell us what to do! Now that we are seeing them in our skies at night, over major cities without any resistance, it is not as far-fetched as you might think! The "Phoenix Lights" is only one major example!

"Lights of varying descriptions were seen by thousands of people between 7:30 pm and 10:30 pm MST, in a space of about 300 miles (480 km), from the Nevada line, through Phoenix, to the edge of Tucson. There were two distinct events involved in the incident: a triangular formation of lights seen to pass over the state, and a series of stationary lights seen in the Phoenix area. Allegedly the United States Air Force identified the second group of lights as flares dropped by A-10 Warthog aircraft that were on training exercises at the Barry Goldwater Range in southwest Arizona. Witnesses claim to have observed a huge carpenter's square-shaped UFO, containing five spherical lights or possibly light-emitting engines. Fife Symington, the governor of Arizona at the time, was one witness to this incident; he later described the object as being "otherworldly."[27]

Why are they here, and what are they doing? We have no way of knowing, because the government won't tell us, but plead ignorance about the whole issue. For me to say they are attempting to control us makes more sense than they are trying to help us. Say if you owned some cattle on a ranch, you don't let them do what they want do you? You pen them in with a heavy-duty fence. You harvest them when you need to as food for your family. You sell them off to stay in business. You maybe milk the cows and eat the steers. You castrate the bulls. They are yours to do with as you want.

WE ARE THE CATTLE OF THE DEMONIC ALIEN OVERLORDS!

No, you are not that way, or so you think. Try to live a day without doing something you know is very wrong, but you do it anyway. Maybe you can hold out for a day

---

[27] Wikipedia, "Phoenix Lights," https://en.wikipedia.org/wiki/Phoenix_Lights Accessed 6.29.2021

or two, but eventually, you will capitulate. You are not in control of your own life. You are under some other control. Yes the Bible calls it sin and everybody knows about that, but this now becomes much more sinister. Back then, it would have been too much for us to understand about Demonic Alien Overlords. It is even now, but all is being revealed as it gets closer to the end.

The Bible tells us how to resist. The Bible tells us the devil is trying to make us sin. We are being tempted by Satan to do evil. We have to defeat him in our own lives to keep from being drawn into a "Hell on Earth." Fall prey to the devil and you will be his. You have to fight with the only weapons that work to defeat this horrible enemy. If you are not a Christian, this will not work at all. You will do whatever the devil wants you to do. If you are a Christian, here is the way to defeat them (the Demonic Alien Overlords) from Scripture:

> **Submit yourselves therefore to God. Resist the devil, and he will flee from you.** James 4:7
>
> **Finally, my brethren, be strong in the Lord, and in the power of his might. Put on the whole armour of God, that ye may be able to stand against the wiles of the devil. For we wrestle not against flesh and blood, but against principalities, against powers, against the rulers of the darkness of this world, against spiritual wickedness in <u>high places</u>.** Ephesians 6: 10 - 18

Notice that the spiritual wickedness is in "high places." Exactly where these Demonic Alien Overlords travel, in high places above us! They are monitoring us, controlling us, implanting us, abducting us, using us for whatever vile purpose they could have, and we are powerless to stop them! Unless we are Christian and we use the Word of God as our weapons to stop their control of us, we are slaves!

Chapter 4
EVIDENCE OF ALIEN-HUMAN HYBRIDS

Since the beginning of time, there has been evidence of genetic manipulation of what the authorities think are aliens mating with humans, but I contend it is demonic exploitation of the human DNA and even demons mating with human women to produce the hybrids. There is ample evidence of skeletons of giants that have been discovered all over the world, but they have been quickly covered up by the people in charge. They don't fit well with the idea that we have evolved and are an unexplained anomaly that is kept hidden away so that Darwin's Theory of Evolution can still be taught, and not the biblical story. If they show the giants, then the scripture I mentioned earlier about giants is shown to be true. Here is one example:

"The allegations stemming from the *American Institution of Alternative Archeology* (AIAA) that the Smithsonian Institution had destroyed thousands of giant human remains during the early 1900's was not taken lightly by the Smithsonian who responded by suing the organization for defamation and trying to damage the reputation of the 168-year old institution.

During the court case, new elements were brought to light as several Smithsonian whistleblowers admitted to the existence of documents that allegedly proved the destruction of tens of thousands of human skeletons reaching between 6 feet and 12 feet in height, a reality mainstream archeology cannot admit to for different reasons, claims AIAA spokesman, James Churchward."[28]

---

[28] From "SMITHSONIAN ADMITS TO DESTRUCTION OF THOUSANDS OF GIANT HUMAN SKELETONS IN EARLY 1900'S," https://worldnewsdailyreport.com/smithsonian-admits-to-

These 12-foot skeletons are not the largest, but skeletons have been discovered to be up to over 42 feet tall![29]

There is always a denial of these skeletons being real, and that pictures are fake, but with the Smithsonian whistle blowers admitting under oath that skeletons of up to 12 feet were destroyed, then who is to say there were not even larger ones destroyed? There are many stories on YouTube of giants, and many newspaper articles that attest to there being giants. But not only giants, but non-human creatures that appear almost human, but are not. Take the case of the elongated skulls in Peru. There are hundreds of skulls that are elongated, and not caused by binding the skull of children. One article in "Ancient Origins Reconstructing the Story of Humanities Past," shows pictures of elongated skulls that DNA testing has shown to be **inhuman** (emphasis added)!

These examples were baffling due to their size and unorthodox structure, and remain so to this day. In fact, DNA testing performed in 2014 perhaps just muddied the water more, when ***a geneticist reported that they have mitochondrial DNA "with mutations unknown in any human, primate, or animal known so far".*** And then later testing at labs in Canada and the USA claimed they were from "the Haplogroup (genetic population group) of H2A, which is found most frequently in Eastern Europe, and at a low frequency in Western Europe."

---

destruction-of-thousands-of-giant-human-skeletons-in-early-1900s/ Accessed 6.30.2021

[29] ECUADOR EXPOSE THE SKELETONS OF AN ANCIENT RACE OF GIANT HUMANS – 7 TIMES BIGGER THAN MODERN HUMANS https://archaeology-world.com/ecuador-expose-the-skeletons-of-an-ancient-race-of-giant-humans-7-times-bigger-than-modern-humans/ Accessed 6.30.2021

So, when it comes to elongated skulls in Peru, there is clearly much more work needed in order to reach any conclusions regarding how or what has caused the misshapen forms and abnormal capacities.[30]

There is also evidence of alien-humanoid skeletons found that are quite small. One such example is the 6-inch-tall Atacama Skeleton found in Chile's Atacama Desert in 2003. The account is absolutely not a hoax of any kind. From the "Paranorms" article "The Atacama Skeleton: Tiny Human Remains?" it recounts the discovery of the skeleton and how it appears to be not human. The article concludes that it is still a human female, but this is not possible as DNA proves otherwise!

The story behind the Atacama Skeleton is a bit strange. A treasure hunter in Chile's Atacama Desert discovered a tiny body in 2003. The body was wrapped in a white cloth and found inside of a leather bag, near a church in the ghost town of La Noria. The discoverer sold the mummified corpse but did not provide many details about the curious find. It's speculated that the treasure hunter was likely a grave robber.[31]

---

[30] From "ANCIENT ORIGINS RECONSTRUCTING THE STORY OF HUMANITIES PAST," "Elongated Skulls Found in Completely New Region of Peru," by Gary Manners, updated August 26,2020. https://www.ancient-origins.net/news-history-archaeology/elongated-skulls-0014172 Accessed 6.30.2021.
[31] From "Paranorms," "The Atacama Skeleton: Tiny Human Remains?" By Ellie Zed. Updated April 2021. https://paranorms.com/the-atacama-skeleton-tiny-alien-remains/ Accessed 6.30.2021

The article has accompanying pictures that show an unbelievable skeleton that is obviously not human, but still, researchers try to say it is. The articles states:

"In the original analysis, Dr. Nolan found that 8 percent of the DNA was unmatchable with human DNA. Clearly, something like that could make headlines! While that was odd, he noted that was due to a degraded sample, not extraterrestrial biology. After further analysis, a more sophisticated process was able to match up to 98 percent of the DNA, according to Nolan."[32]

There is an abundance of evidence that giants and tiny alien-human hybrids existed in the past, and that genetic manipulation has been done in the past, that I can make the statement that since these Demonic Aliens have done it in the past, they will continue to do so in the future. They were slowed down before by God when he sent the Flood in the time of Noah. People don't believe in that anymore, for the most part, but think it is a fairytale. I contend it is true, as is the whole Bible and that the idea that we have somehow evolved from nothing is the fairytale. The excellent work *The Genesis Flood* goes over that in detail and proves the Genesis Flood is true, but in getting back to my contention that Demonic Aliens are controlling us, genetically manipulating us, and even mating with human women, there is evidence for this today. Take the cases where women have given birth to human-alien hybrids or fetuses.

---

[32] From "Paranorms," "The Atacama Skeleton: Tiny Human Remains?" By Ellie Zed. Updated April 2021.
https://paranorms.com/the-atacama-skeleton-tiny-alien-remains/
Accessed 6.30.2021

"I was taken abroad a flying saucer, I was laid out on a metal bed". And you have some body marks from that event. "Yes, I have body marks from the event". And you also said something about a florescent substance…" Yes the substance that was on my body comes from their "Epidermie (don't know the English word for this) that they use to manipulate our body. "They say that they use it as a disinfectant in order not to contaminate us or to get contaminated themselves". "Because they received a lot of irradiations as a direct result of visiting various different planets. Listen, why do they have a need for your tissue? "Mine and others like me are a race that is bio-compatible". "We are the only race that is bio-compatible". "Genetically speaking we are the closest species that is similar to theirs". "We can help them reproduce and re-create a whole race".

You remained pregnant right? "Yes, I had 18 "pregnancy". (I assume it's a sort of a forced abortion). "During the second month they come, and they take the fetus away". And then what happens? You don't feel pregnant anymore? (I'm having trouble hearing this part. Apparently, she says that they come and take her milk from her breast. I don't know how a woman can have milk at that time but whatever).[33]

The article goes into detail about how they have an alien-hybrid human fetus they are doing tests on to discover what it is. An accompanying video shows a non-human fetus that was aborted. The aliens are lying by telling her that they need her genetic tissue to reproduce because they are no longer able to now. What other proof

---

[33] From "Italian Alien Abducted Woman Interviewed About Her Fetus Autopsy," https://www.ufo-blogger.com/2009/07/italian-alien-abducted-woman.html Accessed 7.1.2021

do we need to demonstrate that these are demons and not aliens? These Demonic Alien Overlords are using us to try to produce some type of "Master Race," that will take over our world, and we will then be exterminated out of existence! There are many different accounts of these types of occurrences where an obduction happens, the woman is impregnated and later the fetus is removed by the Demonic Aliens. First of all the government is admitting these UFOs are real, so now we know what their objective is:

WORLD DOMINATION DISGUISED AS BENEVOLENCE FROM DEMONIC ALIEN CREATURES SOMEHOW IN NEED OF OUR HELP!

They are not here to help us, they are here to control and destroy us, and take over the world in the process! They were stopped before by God with the Flood in Genesis. Now they are back again with the same agenda. Ruin God's perfect creation and corrupt what he made! We will no longer be human, but demonically controlled zombies that will do anything the Demonic Alien Overlords ask. When they come out with the fact that if you just take the implant, you can't die, people will be begging to get that implant. Be prepared to resist!

We are all failures. We succumb to sin without any fight at all. I am one of the worst offenders and now I hardly try to resist. I submit to the temptations they "shoot" down from above. How else are they doing it? They inhabit our consciousness and rule us like we are their slaves. The only way to fight them off is to become a Christian and prepare for battle! You have to read the Bible daily, attend church whenever you can, pray for forgiveness, and work to defeat them in your own life.

Then you can tell others how to win the battle. When I first became a Christian, on January 20, 1980, I started to win. Now they have turned up the heat. I wasn't any threat to them back then. Now they make sure they are controlling me better, by increasing temptations and monitoring me to make sure I "get in line!" I hate to tell the truth when it comes to my failures, but we are all in the same boat. None of us is really winning. Jesus has defeated them, though! Trust in him and you will win!

Chapter 5
HOW TO WIN THE BATTLE

I need to tell you know how to defeat them. I know most people don't want to hear about how to become a Christian and all of that. Most won't believe in Jesus now, because look at what we know about everything. We evolved and were not created, you say. We have aliens showing how this happened. Some scientists even say the Earth was "seeded" by aliens. But, just when you thought it was safe to not believe in God, these Demonic Alien Overlords are being revealed. Just like in the movie "Jaws" when they said, "Just when you thought it was safe to go in the water," it isn't safe to walk around without being a Christian These UFOs that are flying in our skies are the demons the Bible talks about, so please believe what I am going to tell you now. The standard Gospel of Jesus is still the only way to survive. You are either with God and saved forever, or you are without hope and controlled by Demonic Alien Overlords!

The Plan of Salvation from the King James Version of the Holy Bible in the Public Domain

The Bible makes it clear there is only one way to get to Heaven, and that way is by trusting in Jesus Christ alone for your salvation, repenting of your sins and asking him **(praying out loud)** to come into your life and save you from your sins:

"These things have I written unto you that believe on the name of the Son of God; that ye may know that ye have eternal life, and that ye may believe on the name of the Son of God." (1 John 5:13)[34]

---

[34] Holy Bible, King James Version, in public domain, 1 John 5:13

Jesus said: "I am the way, the truth and the life: no man cometh unto the Father but by me." (John14:6)[35]

<u>There is nothing you can do on your own that will save you, all of the good works you do are worthless.</u>

"For by grace are you saved through faith, and that not of yourselves: it is the gift of God: Not of works, lest any man should boast." (Ephesians 2:8-9)[36]

Trust Jesus Christ alone for your salvation! Here is the only way you can get to Heaven:

<u>You first have to admit you are a sinner and be aware that you cannot save yourself.</u>

"For all have sinned and come short of the glory of God;" (Romans 3:23)[37]

"Wherefore, as by one man sin entered into the world, and death by sin: and so death passed upon all men, for that all have sinned." (Romans 5:12)[38]

"If we say that we have not sinned, we make him a liar, and his word is not in us." (1 John 1:10)[39]

<u>Repent of your sins, turn away from them and stop doing them.</u>

Jesus said: "I tell you, Nay: but, except ye repent, ye shall all likewise perish." (Luke 13:5)[40]

---

[35] Holy Bible, King James Version, in public domain, John 14:6.

[36] Holy Bible, King James Version, in public domain, Ephesians 2:8-9.

[37] Holy Bible, King James Version, in public domain, Romans 3:23.

[38] Holy Bible, King James Version, in public domain, Romans 5:12.

[39] Holy Bible, King James Version, in public domain, 1st John 1:10.

[40] Holy Bible, King James Version, in public domain, Luke 13:15.

"And the times of this ignorance God winked at: but now commandeth all men everywhere to repent:" (Acts 17:30)[41]

<u>Believe in Jesus Christ alone for your Salvation because he alone died for you, was buried, and rose from the dead.</u>

"For God so loved the world, that he gave his only begotten Son, that whosoever believeth in him should not perish, but have everlasting life." (John 3:16)[42]

"But God commendeth his love toward us, in that, while we were yet sinners, Christ died for us." (Romans 5:8)[43]

"That if thou shalt confess with thy mouth the Lord Jesus, and shalt believe in thine heart that God hath raise him from the dead, thou shat be saved." (Romans 10:9)[44]

"Behold, I stand at the door, and knock: if any man hear my voice, and open the door, I will come in to him, and will sup with him, and he with me." (Revelation 3:20)[45]

<u>I "believe" in Jesus is not enough!</u>

"Thou believest that there is one God; thou doest well: the devils also believe, and tremble." (James 2:19)[46]

<u>Heaven is a free gift but must be received.</u>

---

[41] Holy Bible, King James Version, in public domain, Acts 17:30.

[42] Holy Bible, King James Version, in public domain, John 3:16.

[43] Holy Bible, King James Version, in public domain, Romans 5:8.

[44] Holy Bible, King James Version, in public domain, Romans 10:9.

[45] Holy Bible, King James Version, in public domain, Revelations 3:20.

[46] Holy Bible, King James Version, in public domain, James 2:19.

"But as many as received him, to them gave he power to become the sons of God, even to them that believe on his name:" (John 1:12)[47]

<u>You must pray the Prayer of Salvation out loud.</u>[48]

"For with the heart man believeth unto righteousness, and with the mouth confession is made unto salvation." (Romans 10:10)[49]

"For whosoever shall call upon the name of the Lord shall be saved." (Romans 10:13)[50]

<u>Here is a simple prayer that will save you, if you believe, are sincere and pray it out loud:</u>

"Dear Jesus, I repent of my sin and ask you to come into my heart and save me from my sin. I believe in you and believe that you died on the cross and rose from the dead to save me from my sin. I make you Lord and Savior of my life and I invite you to come into my heart and save me! Help me to live for you. In Jesus' name, Amen."

<u>Once you are saved, your Salvation is forever, and you cannot lose it!</u>

"And I give unto them eternal life; and they shall never perish, neither shall any man pluck them out of my hand." (John 10:28)[51]

---

[47] Holy Bible, King James Version, in public domain, John 1:12.

[48] The Message of Salvation, Holy Bible, King James Version, in public domain.

[49] Holy Bible, King James Version, in public domain, Romans 10:10.

[50] Holy Bible, King James Version, in public domain, Romans 10:13.

[51] Holy Bible, King James Version, in public domain, John 10:28.

"All that the Father giveth me shall come to me; and him that cometh to me I will in no wise cast out." (John 6:37)[52]

<u>You have forgiveness of sin when you confess them once you are saved.</u>

"As far as the east is from the west, so far hath he removed our transgressions from us." (Psalm 103:12) [53]

"He will turn again, he will have compassion upon us; he will subdue our iniquities; and thou wilt cast all their sins into the depths of the sea." (Micah 7:19)[54]

"If we confess our sins, he is faithful and just to forgive us our sins, and to cleanse us from all unrighteousness." (1 John 1:9)[55]

If you have prayed the Prayer of Salvation begin going to church and tell someone that you have accepted Jesus. Publicly profess that you have become a Christian at church when they have an alter call. Be baptized as soon as possible.

Most people hope there is another way. That the Bible is fiction, and these aliens are from another planet like on "Star Trek," but that is not the case. It is all being revealed to be true now, more than ever before.

---

[52] Holy Bible, King James Version, in public domain, John 6:37.
[53] Holy Bible, King James Version, in public domain, Psalm 103:12.
[54] Holy Bible, King James Version, in public domain, Micah 7:19.
[55] Holy Bible, King James Version, in public domain, 1 John 1:9.

Chapter 6

ALL HELL WILL BREAK LOOSE!

Things continually get worse by the day. There are riots in the streets, people get shot and killed for no reason in the major cities. Crime goes unpunished. It is like it will be just before the Rapture. Then these Demonic Alien Overlords step in and say they are here to "save" us. Billions of people disappear worldwide, and they say they all were "transported" to their home planet, wherever that is! Like it was at the beginning, they take credit for everything God has done. They will say they created us and now they are here to lead us to a utopia of "Heaven on Earth." It might just be that for those who take their implant for the first three and a half years of the Tribulation. People who don't take the implant get rounded up and put in the FEMA camps and are later executed by guillotine. Everything is in place for all this to happen, just like the Bible says. Do you really want to take the chance and think it is not true? Even right now it is so bad, you can't walk out of your house without your mask. You are likely to be killed by a crazed gunman. You are in jeopardy of being arrested by the police for looking at them wrong, and if you want to protest our government in any way, you are thrown into jail and never seen or heard from again. The United States is becoming like Communist China at an unbelievable rate. The Constitution is ignored, and we are being manipulated, by the real entities in control of everything, the Demonic Alien Overlords!

You can say I'm wrong all day long, but once the Rapture happens you will see the truth if you are still here. Just remember my warnings, and please get saved before it happens, or else you will be condemned to Hell when you take the implant. You will have no choice if you want to eat. Take the implant or starve to death, or be beheaded by guillotine! Few will be able to hold out until the end.

When they are finally completely disclosed, you will not be able to even say anything bad about them. Even today, if you want to say something bad about a candidate for office, for instance, or another person that the powers that be don't like what you say on Twitter, YouTube, or any of the so-called free message boards, they will kick you off of them. They call it misinformation, even though free speech can be wrong and still be able to be said. But once these creatures are revealed, you will be arrested and beheaded for saying the things I have said in this book. I have time now before they are revealed. After they are disclosed completely, they will be our "Masters," and we will be their slaves. If I were even now putting this idea out there, on the Internet or on Twitter, or some other similar place, it might be censored. I have no proof of what I am saying, but there is no absolute truth to those that don't accept it. Even when something is absolutely true, like $2 + 2 = 4$, the deniers will swarm in and say it is false. There is a whole group of people claiming that the world is flat, but they don't get censored. Neither should I from what I am saying because right now, I can make the claim that demons are not even real, but the truth is they are! They are the aliens that inhabit those UFOs we see everywhere! How many times do I have to repeat myself?

SCIENTIST CLAIMS UFOS ARE DEMONIC

In "VYC America", there is the claim that UFOs are satanic. I have seen other quotes from scientists, and Dr. Don DeYoung states:

"…a scientist and a born again Christian explaining the phenomenon of UFO's and saying that these UFO's could well be satanic creatures, evil angels."[56]

Others, not just me, hold this same view. Skywatch TV claims their origin is satanic.

Their belief is that the UFO phenomenon is 100 percent real, but that its origin is satanic. As they see it, the "aliens are visiting us" angle is a camouflage to allow Satan—if he really exists, of course—to get his grips into us. They believe that the presence of the UFO phenomenon, right now, and since 1947, is the commencement of an "end times" scenario, where the "E.T. landing" angle may be played out to con the world into believing the "Grays" are friendly extraterrestrials. The CE (Church of England) believes that belief and prayer, rather than weapons, can hold off this infiltration and invasion.[57]

Now that the government has revealed that UFOs and UAPs are real, they have stopped short of saying we have any idea who the occupants of these crafts are. They are for once telling the truth, they don't come from outer space, but from here. But they are omitting the fact they are demons. They have been observed traveling under water as well as in the air. They have bases right here on the Earth and under the oceans. That is what Admiral Byrd discovered when he went to the Artic.

---

[56] From "VYC America, Christian Information Radio. TV Online," by Jimmy DeYoung. 6.6.2019
https://www.vcyamerica.org/prophecy-today/2019/06/06/a-scientist-says-that-ufos-could-be-satanic-creatures-2/
Accessed 7.2.2021

[57] "The Coming Great Deception – Part 16: Portals, Occult Magic, and the Collins Elite," by SkyWatch Editor. 6.1.2021. https://www.skywatchtv.com/2021/06/01/deception16/ Accessed 7.2.2021

Whenever they get close to telling who they really are, they stop short. I believe that John F. Kennedy was assassinated because he wanted to reveal that they exist. The Deep State couldn't let him give a speech saying we have alien technology, and that we have live aliens that are helping us. They also couldn't let Trump disclose the truth about them. He started the Space Force to combat these creatures, although he is misguided because they are from here. He had to be defeated, so the Deep State saw to it that he lost the election. At least this time they didn't assassinate him. They came up with a bogus report that we just don't know what these UFOs are. These Demonic Aliens have been revealed to the government since at least from the time of the Roswell Crash and the rumors are that one of them survived the crash. Eisenhower met with them and signed a treaty with them, it is rumored. We get their technology, and they get genetic material so they can reproduce. We assist them to gather it, hence the black helicopters when there are cattle mutilations.

I can't get real evidence for any of this, but it just seems too easy for Oswald to shoot Kennedy like he supposedly did. It is easy to see how there had to be more than one person involved in it. Why keep the whole report secret until now about who shot him? Trump also wanted to disclose that. So now he had to be defeated, no matter how it was done, so the Deep State engineered the loss. They used software called "Hammer" and "Scorecard" developed to interfere in other elections for foreign candidates and used it here to make sure Biden was elected. I heard this only a couple of times on YouTube, and then it was removed. YouTube videos will now state the software programs of Hammer and Scorecard are a hoax. Whatever the answer is, just realize the time is coming when these alien creatures will be fully disclosed. Remember, they are demons and not aliens. Do not take their implant and resist with all your might. Become a Christian today to make sure you are not left behind to go through the Great Tribulation. Before the aliens are disclosed, I believe, the Rapture will occur. I don't know the date, but very soon. I only hope I get this book published and in bookstores before any of that happens.

Theologians will tell you this Great Tribulation will last for seven years. The first three and a half years won't be that bad. There will be restrictions on who can buy or sell, and you have to get the implant, but for the first part of that time you might be able to find some food, but it will run out completely, hence the terrible time of the last half of it where almost no one will be able to survive unless you take the implant. When you see your children and your wife starving, you won't hold out against the Demonic Aliens. You might hold out if it were just you, but if you have a family and they are with you, you won't be able to sit by and watch them starve. You will take the implant and tell them to do it also, so you can get something to eat. Mark this page. I told you so!

Take the implant and you will be changed from human to a human-alien hybrid. You will follow blindly behind the demands of your Demonic Alien Overlords. Other humanoid beings will catch you, or turn you in if you don't want to get the implant. They will be like zombies that just do what the implant tells them to do. It may not be the implant someone on Earth made that is the one that is the Mark of the Beast. It will be one the aliens made, most likely, and they will force everyone to get it. It is what the Bible talks about in Revelation. People take the mark so they can buy or sell. It will be administered by the demands of our Demonic Alien Overlords, I believe, not just by men. They demand people get it, and they do. They arrest people who don't and throw them in FEMA camps. You can shoot the people with the implant, and they are healed from the shot. They are the slaves of demons. The only way to kill them is to cut their heads off, but that is very difficult. Here is what the Bible says about that mark:

> **16 And he causeth all, both small and great, rich and poor, free and bond, to receive a mark in their right hand, or in their foreheads:**
> **17 And that no man might buy or sell, save he that had the mark, or the name of the beast, or the number of his name.**
> **18 Here is wisdom. Let him that hath understanding count the number of the beast: for it is the number of a man; and his number is Six hundred threescore and six.** Revelation 13: 16 - 18

They are already starting to implant people. The evidence is clear that the Bible is true. We now have the technology to make sure people don't get to eat unless they have the implant. The Bible is coming true right before our eyes! The Bible has also this strange passage:

> **And said to the mountains and rocks, Fall on us, and hide us from the face of him that sitteth on the throne, and from the wrath of the Lamb:** Revelation 6:16

So, you can't die either, but it won't be painless. And once you get that implant you will be condemned to Hell. In the end, people who take the implant, suffer eternal damnation and separation from God. The Christians are raptured out before all this happens; many biblical scholars believe. Once you take the mark, you will think everything is great! No more sickness, no more death, but you missed the part where you start to get sores all over your body from the implant!

<blockquote>
⁹ And men were scorched with great heat, and blasphemed the name of God, which hath power over these plagues: and they repented not to give him glory.

¹⁰ And the fifth angel poured out his vial upon the seat of the beast; and his kingdom was full of darkness; and they gnawed their tongues for pain,

¹¹ And blasphemed the God of Heaven because of their pains and their sores, and repented not of their deeds. Revelation 16:9 – 11.
</blockquote>

These Demonic Aliens convince you it was all caused by God, not them and that he is the reason you are in constant pain but can't die or get relief. You are still blinded to the fact that it is the Demonic Alien Overlords that have done this to you, and not God. Sores and constant pain and agony caused by taking the implant which they told you would give you eternal life! You have a choice to make now before you have to take it. Repent and be saved, and get taken out by the Rapture before Jesus returns, and then you will have eternal life and eternal happiness and joy. Let's see, agony and pain and torment in Hell with the devil, or eternal life, happiness and joy with God! Which one is better? Obviously, you know which one is better, but people still don't want to believe it. From the time you were a child, these Demonic Alien Overlords have been drumming into your head that God does not exist. We evolved from the goo in the bottom of a cesspool in a septic tank or something. Either that or when these demons are revealed, they will take the credit for the creation. They were the ones who "seeded" the Earth with life and man popped out! We have evolved over the millions of years of evolutionary change, and now we are finally our own gods! But we have no purpose for our life, no meaning. We have to serve our Demonic Alien Overlords and do everything they say! When we die, we go into the dirt and are eaten by worms!

With these demons in charge of my life, I need to take drugs, rob and cheat, rape and murder, because when I die it is all over. With God, we have a purpose, to know God and serve him, we go to Heaven and have eternal life, joy, and happiness! With that hope, I can do anything I want; write this book, invent something, become an industrialist, whatever I want. What a difference a belief makes!

Still, people refuse to believe. It is the constant drumming of lies into your head! I am now becoming convinced that it may just be, these alien spacecraft; hoovering over us, sending "rays" down to convince us to do what they want! Maybe it is just that simple, they cause us to sin by the demonic oppression that reaches our minds and controls them. We have a way to battle it, though! Again, once you become a Christian, you also receive the Holy Spirit. That will help you to overcome the devil. When you read the Word of God, you are strengthened against the sin desire from the "demonic oppression" being shot down on us by these demons. The Bible does not describe sin like this exactly, but how sin fills our minds and hearts is up for discussion. We choose to do it, or we don't, and we are constantly oppressed by the devil, and what better way for him to do it, is from alien spacecraft hoovering above us unseen and unknown, and able to shoot those rays down to control us, clandestinely. The waves fill our brains with the desire to sin, and we do it.

Some people aren't bothered too much by these oppressions, thoughts, and desires to sin. When I was a boy, I was kept in line by my parents that loved me. I was also no threat to the plan of Satan to rule the world and destroy God's creation, so I was not oppressed that much by the devil and these aliens. Some children are very susceptible to oppression by demonic spirits because they allow the aliens to enter their minds and control them. Adults have the same ability to either give into the oppression or repress it. We are still doing the bidding of these aliens because we give in when we do wrong. We don't give into their bidding, sometimes out of fear, other times out of fear of rejection by our peers, and sometimes out of stubbornness. We are still being manipulated and controlled by these demonic spirits, but sometimes we can resist some. We have a little will to keep from doing what they want, but when once the Rapture happens, we will lose our ability to resist. The Holy Spirit in some people restrains us from doing some bad things because we still want to receive acceptance by others, but that does not stop many. Some are going to give in out of peer pressure to rob, kill and rape because they are members of a gang, which encourages them to do such evil things. The Demonic Alien Overlords are more in control of us when we let them. If we resist or are a Christian, allowed to be controlled by the Holy Spirit, we will win the battle. If not, we will always lose.

I have no proof for any of what I am saying about these supposed aliens, that they are out there and controlling us. I have no proof that they are the demons that were here at the time of Jesus. But what better explanation is there? These creatures show up in what appears to be alien spacecraft and we are supposed to believe they are from some faraway galaxy, and they traveled here to help us. Why would they do that? It isn't *Star Trek, E.T.,* or *Starman.* It's *Independence Day,* and we had better be ready to fight!

Chapter 7
THE TIME IS AT HAND!

So, when is all this supposed to happen? The government was supposed to disclose about these UFOs or UAPs in June of 2021. They told us nothing, of course! They have all the information but are still lying, again because the Demonic Alien Overlords that are actually controlling everything, don't want them to tell us. They won't be revealed until after the Rapture of the church, I am convinced. Any enemy that wants to conquer a foe, does not broadcast its plans until the moment of the surprise attack. They will wait until the last minute to strike! So, these Demonic Alien Overlords will arrive **after** the Rapture, to step in to claim they are the reason and not God! The men, women, and children that are gone went to their planet, and you can too, just take the implant! Any enemy trying to defeat a foe will also lie, in order to cause confusion, so they can defeat the unprepared and unsuspecting adversary. We will not go to their planet, again, their planet is here! So, when does the Rapture of the church happen? Sooner than you think!

There are signs all over it will happen any day now. We have the warnings of others about when we can expect it to happen. The Nation of Israel becoming a nation again is a key sign that the time is now! Jesus states:

> **28** Now learn a parable of the fig tree; When her branch is yet tender, and putteth forth leaves, ye know that summer is near:
> **29** So ye in like manner, when ye shall see these things come to pass, know that it is nigh, even at the doors.
> **30** Verily I say unto you, that this generation shall not pass, till all these things be done. Mark 13: 28 - 30

Scholars may disagree that the "fig tree" here represents the Nation of Israel, but if it does, then Jesus could return at any time now. Israel became a nation on May 14, 1948.[58] When it talks about a generation, some have thought it means 40 years. If it means everyone that was born in that time, maybe it means up to 120 years or more before everyone dies out that was alive at that time, then we may have a little while yet, but it could happen anytime. If it is 120 years, then by May 14, 2068, but it may be earlier. I am not setting any dates, just surmising what might be a possibility here. Everything that has to happen has already happened that needs to occur in order for the Rapture to take place today according to most biblical scholars. So, with that, we know it will not be long. Maybe before I finish…! Just kidding, it can happen any time now!

There are other signs happening that state it is soon. One highly respected Rabbi, Yitzhak Kaduri, shortly before his death, reportedly saw a vision of Jesus, and that

---

[58] From "World Atlas," "When Did Israel Become a Nation?" https://www.worldatlas.com/articles/when-did-israel-become-a-nation.html#:~:text=Israel%20is%20the%20biblical%20Holy%20Land%20in%20the,declaration%20of%20the%20establishment%20of%20the%20Jewish%20State. Accessed 7.5.2021

he would return soon.[59] Although we don't know how soon, it could be today! Then once all this happens, that is when all the Demonic Alien deception really starts! Today they are beginning the brainwashing, to get billions to believe they are aliens. Then they spring themselves on us, and billions will believe.

Someone once said, that if the devil just showed up all at once and said he wanted to take you to Hell with him, you wouldn't take his offer, of course. But if he slowly raises the temperature, (increases our acceptance of sin) like how a frog in a pot never jumps out, and will be boiled to death in a pot, he has you. Today he is raising the temperature, on a gradual level, but getting hotter by the day, and we are the frogs! Jump out now! Accept Jesus **_now_** before it is too late, and you are lost forever and condemned to Hell!

Back when I was a child, there were no discussions about the perversion we see today in our societies. We were shielded from the perversion of being gay, or from a child molester coming into your room at night. It just wasn't talked about. Things like this rarely happened and people that had perverted lifestyles were shunned. Today, our whole nation has been reduced to a cesspool of perversion. Stuff like pornography, child molestation (coming soon to be allowed by a courtroom near you), drag queens telling stories in school, and evil perversions, were all kept behind a closed closet door somewhere where we wouldn't be subjected to these evil things. Now we are expected to just accept them as an "alternate lifestyle" we have to accept. It wouldn't have been accepted then and shouldn't be accepted now. But we have been

---

[59] <u>Finding Proof of Jesus,</u> by James L. Kearns. WestBow Press 3.7.2016

"heated up" like the frog in the pot, and soon we will all be cooked!

These things are all an indication that the Rapture will happen sooner than most of us think. It is getting so bad now, that it must happen soon. Covid-19 was a precursor of what is to come. It wasn't the true end-time but showing us what it will be like then. The Bible also talks about how there will be pestilences in the last days. Consider this Scripture:

> **[7] For nation shall rise against nation, and kingdom against kingdom: and there shall be famines, and <u>pestilences,</u> and earthquakes, in divers places.** Matthew 24:7

We have a situation here when we will be forced to take the implant, to combat the disease that is thrust upon us or be considered an enemy of the state. After the Rapture, it will happen again, and this time you will have no choice but to get the vaccine, which will also include the implant, or you will be rounded up and put in the FEMA camp. Our "alien friends" will decree it, and you will have no choice but to comply or die! This time you had the choice of taking the vaccine or not, but you could not go into a business anywhere without a mask, or you would be arrested. Next time it will be a forced vaccine and failure to comply will get you arrested on site.

It is getting closer and closer by the day. As I write this, there may be only months to go before the Rapture occurs or even days, but no one knows when. I am compelled to write this "warning" because the time is truly short. If it were entitled <u>The Rapture</u>, no one outside of the Christian community would even consider reading it, but since it contains UFOs and aliens, maybe some people from the secular community will consider it as a possible explanation for what is happening now with UFOs.

They have been trying to get a foothold on the world from the beginning. It has happened before in the past, specifically with the Incas in Peru. The majority of the elongated skulls are clearly not human, no matter what they say. That was the genetic manipulation by the lizard-like aliens, most likely. But who knows what evil Demonic Alien creature mated with humans to produce such a monstrosity? Once the Rapture happens, the people that are left that take the mark will be changed into a creature that has no will to resist any of the alien demands. They won't even know they are controlled but will follow blindly, like sheep to the slaughter. They will then blame God for their sores, not the Demonic Alien Overlords who are really responsible for their predicament.

Many people will make all kinds of excuses rather than to believe in God, and when the Rapture happens, they will now have one, they think, that the Aliens, not God, did it. Satan always tries to take all the credit for everything. Not many truly believe the Bible, but substitute evolution as the answer to why we are here, and that plays right into Satan's hands. When confronted with the possibility of the return of Jesus, the majority of people that aren't saved give this excuse:

> **And saying, Where is the promise of his coming? for since the fathers fell asleep, all things continue as they were from the beginning of the creation.** 2 Peter 3:4

They choose not to believe. They don't believe in God but can be persuaded to believe in aliens that will masquerade as "God" because it is easier to believe in that than a God that has rules and regulations, and wants us to know him. Since it is aliens, all bets are off. We don't have to obey the Ten Commandments, but what they don't know is that once the aliens take over, there will be hellish rules we have to follow, or else. Right now, we have a period of grace. After that, grace will end.

Demonic Aliens or demons aren't involved with the Rapture at all, except that after it happens, then they will be in charge of the world until the end of the Tribulation, which will last for seven years. Think they are not going to make you do what they say? Think again! Aliens from supposedly another planet land at the White House and they roll out the red carpet for them. They are so far above us technologically that the military has no chance to defeat them. They are just here to help us! They are here to help us into **_Hell_**! They will proclaim there is no God, and that they "created" us by evolution and seeding the Earth. Billions will believe this lie. It will be too late for them because the Rapture has already happened, and billions have just disappeared. Now comes the real Hell on Earth!

The Demonic Alien Overlords are waiting for the right time to strike. They can't control Christians, and people like me would be telling everyone they are demons and not aliens. That is why I believe the Rapture happens before they disclose themselves to the world. June of 2021 when the government was supposed to tell us everything was just a ruse. We are being invaded even now, and their goal is total world domination! That wouldn't be as bad, but the deep state is helping them with that goal. The people in charge and collaborating with them expect to be put in leadership roles with this New World Order when the Demonic Alien Overlords rule the world. It is all about power and control, and we are powerless to stop them. Except you become a Christian before it is too late, you will be swept into this Hell they bring on the Earth. While some pastors have hinted you can be saved after the Rapture, don't count on it. The time is now or never, because then you will be unable to resist, you starve or have your head cut off. You will have to watch your other family members (mother, father, sister, brother, or children) get their heads cut off. You have to listen to your children crying constantly for food. You will take the implant to ease their suffering as well as your own. Few, if any will be able to hold out until the end of the Tribulation. Don't think you could pass this test!

Here is the scenario that I believe will happen. The Rapture of the church can happen at any time now. Trust me, everything that needs to happen has already happened. Then Israel signs a peace treaty with the Antichrist. Why not one of the humanoid-alien hybrid creatures that were developed by these Demonic Alien Overlords? It would make sense that a creature that was part alien and part human could fool the Jewish people into believing that he was their Messiah. He has to be a "man" but how would you know if he is part alien or not? He most likely is alive today, in order for the Rapture to happen now. We have no idea who it is, though, and if someone that was half-alien and half-human shows up on the scene, which would be a real eye-opener! The book of Daniel describes the covenant made with Israel and the Antichrist. It ushers in what they think is peace, but it is the Tribulation.

> **27 And he shall confirm the covenant with many for one week: and in the midst of the week he shall cause the sacrifice and the oblation to cease, and for the overspreading of abominations he shall make it desolate, even until the consummation, and that determined shall be poured upon the desolate.**
> Daniel 9:27

One week is seven years and during this time the Temple is rebuilt, and everything seems fine for three and a half years, then comes the Abomination of Desolation where he declares himself to be God. It can happen at any time with all the turmoil happening there, that the Rapture ushers all this in. Right now, they can't rebuild their temple. After the "turmoil" of the Rapture, it would be possible, when the Demonic Alien Overlords are directing it to be done. Are the Muslim countries going to fight against them? I think they would give up when they see Demonic Aliens in their spacecraft land all over the Earth.

Now you really think I am crazy! Where in the Bible does it say anything about aliens? It doesn't and that is just the problem. Again, we are not talking about aliens but demons. It is full of demonic references, and angels, but no aliens.

ONCE AGAIN, THESE CREATURES ARE NOT ALIENS, BUT DEMONS.

The world will be misled into believing they are aliens. That is why they will so easily follow them. Here is the passage that describes how people will be led astray:

The world will worship this "beast." What better describes a beast than some type of human-alien hybrid? I think I may be on to something here! Perhaps it is just a normal man, like Trump or Obama, but that seems very unlikely. No one like that could ever really get people to worship them, could they? Maybe some people might be naive enough, but the whole world? An alien-human hybrid is a shoo-in for the world to worship. They are looking for a "savior" just not Jesus Christ. They don't want him. They want someone else, anyone else will do, and when the demons reveal him, he will appear to be just what the doctor ordered. Maybe he does show up just before the Rapture. The Bible says he will even almost fool some of the elect (Christians), but it could be talking about those that will hold out until the end of the

Tribulation and accept Christ during it if such a thing is possible. One passage reads:

> **For false Christs and false prophets shall rise, and shall shew signs and wonders, to seduce, if it were possible, even the elect.** Mark 13:22

Another passage reads:

> **And except that the Lord had shortened those days, no flesh should be saved: but for the elect's sake, whom he hath chosen, he hath shortened the days.** Mark 13:20

It could also mean the Jewish people will not be fooled by him. Right now, no one I know fits the bill of being worshiped by the world, but a non-human-alien hybrid, a man that is part demon, now that would be someone the world would worship. Whatever it means, whoever it is, when the Demonic Alien Overlords are actually revealed, it will shake the world to its core. What else are people going to think? *Our Saviors are here to deliver us*, but they will be ***demons to dominate us like never before.***

I wrote another book, The Chronicles of the Revelation a few years back. It was fiction about what it might be like after the Rapture if you're not saved. I only wish this was fiction, but I am convinced it's not. These UFOs or UAPs are declared now by the US Government to be real, so now it is going to happen just like I am telling you. One day they will be ready, maybe next year to tell us the whole truth, and then it will happen before our eyes. They will reveal that "aliens" are piloting these crafts, but as I have said before, they are demons. Billions will wander after the Beast to worship these creatures from another planet, and so begins the downfall of the world. God the Father, Jesus Christ, and the Holy Spirt, are the only ones who should be worshipped, not some

Demonic Alien Overlord. When they are disclosed, they will be worshiped by those who are deceived. They won't even know they are doing wrong.

We think we are so intelligent and know everything. We are told over and over by astronomers that somehow there has to be other life out there in such a vast universe. For years there have been all kinds of signs that aliens are visiting us from some other world. It all comes down to if you believe what God said or what man said. The Bible says man was created in the image of God. It makes no mention of aliens on another planet. If there were real aliens out there, and God is the author of the Bible, it would make sense for him to tell us about them also. He only mentions demons, angels, and man. Not demons, angels, aliens, and man. Therefore, it is a safe bet there are no "aliens" but only angels and demons apart from us. If evolution is true, why didn't the moon have life evolve on it, just like us? Mars, and all of the other planets as well? God shouldn't have left out the rest of the story. That would be akin to a lie. Not telling the full truth is not telling the truth. Since God is not a liar, there are no aliens, and those that masquerade as such, are liars, demons, and absolute evil.

I AM GOING TO REPEAT MYSELF AGAIN: THESE ALIENS ARE DEMONS.

Just the good old-fashioned satanic evil beings that were from the beginning. It was a third of the angels that rebelled according to scripture, and these are those creatures, no doubt in my mind!

Chapter 8
PARANORMAL PLACES

As it gets closer to the time when Jesus returns in the Rapture, there will be more and more incidents of UFO activity and demonic attacks. The demons know their time is shorter now than ever before, so they are turning up the heat, to try to win as many battles as possible now before they are defeated forever. One place where we find that an abundance of UFO and demonic occurrences are happening is a place called Skinwalker Ranch. It is a ranch in Utah that has many strange occurrences. People that have owned this property have experienced orbs, UFO activity, strange portals, cattle mutilations, and strange creatures appearing and acting strange, like no other place in America or the world.[60] A "Skin-walker" refers to the term, "In Navajo culture, a skin-walker (Navajo: *yee naaldlooshii*) is a type of harmful witch who has the ability to turn into, possess, or disguise themselves as an animal. The term is never used for healers." [61] Supposedly, I heard somewhere, there was a curse put on the property by a Navaho witch, and that is the reason for the strange occurrences. I have a different theory. This is one of the many "bases" the Demonic Aliens are using to invade us and tempt us to do their evil bidding. The many instances investigated are very numerous. Wikipedia states:

"The ranch, located in west Uintah County bordering the Ute Indian Reservation, was popularly dubbed the **UFO ranch** due to its ostensible 50-year history of odd events said to have taken place there. According to

---

[60] From "Skinwalker Ranch," *Wikipedia,* the free encyclopedia, https://en.wikipedia.org/wiki/Skinwalker_Ranch Accessed 7.6.2021
[61] From "Skin-walker," *Wikipedia,* the free encyclopedia, https://en.wikipedia.org/wiki/Skin-walker Accessed 7.6.2021

Kelleher and Knapp, they saw or investigated evidence of close to 100 incidents that include vanishing and mutilated cattle, sightings of unidentified flying objects or orbs, large animals with piercing red eyes that they say were unscathed when struck by bullets, and invisible objects emitting destructive magnetic fields. Among those involved were retired US Army Colonel John B. Alexander who characterized the NIDSci effort as an attempt to get hard data using a "standard scientific approach". However, the investigators admitted to "difficulty obtaining evidence consistent with scientific publication."[62]

These occurrences are among the many recent occurrences of UFO activity, concentrated on certain areas of the Earth. This would lend credence to the fact that these supposed UFOs are not from some faraway planet but are right here and have been here since recorded time. If aliens are traveling from other planets to the Earth, we should see some indication of that from observable space observation, but it is almost always confined to the Earth itself. By that, I mean, when they observe outer space, and when we look at other planets, we should see evidence of alien life, but as yet we see none. The farther a planet is away from us, in light years, would seem to limit the ability of any alien race to visit us. Now they may have overcome that obstacle technologically, but, if so, why are they visiting this little speck in the universe? We have nothing to offer them, but they would have much to offer us. Always, whenever other people discover another race of humans living somewhere, they think of ways to exploit them, not offer goodwill to them. These

---

[62] From "Skin-walker," *Wikipedia*, the free encyclopedia, https://en.wikipedia.org/wiki/Skin-walker Accessed 7.6.2021

creatures, if they are aliens, would have no reason to be benevolent, but rather, and much more likely, have the ability and the nature to exploit and conquer us, and just take from us whatever they need. Being naive enough to think otherwise is really stupid. All the alien movies I've ever seen don't expect that end. They are always cast as evil, which they would be. That is why even if I'm dead wrong and these creatures are just some alien race here from the planet X, they are here to exploit and conquer us, take what they need, and leave. There can be no other conclusion. But I am giving, conceivably good evidence here, to show they are actually the demons here from the rebellion of Satan at the very beginning.

If this were a baseball game against the supposed aliens, the score would be zero for the aliens and an unlimited number for demons. There is no indication of aliens, but plenty for demons. The Bible talks about them, in our daily lives we are affected by them and drawn into sin, but for aliens, there is no conclusive evidence. Just because they travel in strange spacecraft doesn't make them aliens. When the Bible talks about angels it states:

> **[4] What is man, that thou art mindful of him? and the son of man, that thou visitest him?**
> **[5] For thou hast made him a little lower than the angels, and hast crowned him with glory and honour.** Psalm 8: 4 - 5

Again the Bible declares that:

> **The _chariots_ of God are twenty thousand, even thousands of angels: the Lord is among them, as in Sinai, in the holy place.** Psalm 68:17

Notice the word "chariots." What is a better description of the spacecraft that the Demonic Aliens travel in? The Bible tells us what these beings are, fallen angels that travel to and fro on the Earth in their spacecraft (chariots) as it says in Job 7. It is pretty logical to now assume these are the fallen angels the Bible talks about. If there were truly aliens somewhere, the Bible would have told us. We are "just a little lower than them," so we don't yet have their spacecraft, but we are able to be saved and they are not. Please drum this fact into your head, ***the UFOs are demonically controlled craft, which are not from another planet, but from right here on the Earth. They have hidden bases on the Earth, under the oceans, and even in lakes all over the world.***

Many different places on the Earth are exhibits of the Demonic Aliens influencing people in the past. The Nazca Lines in Peru are an indication of Demonic Alien Overlords manipulating a people because these lines are only really visible as to what they are from the air. While it is possible to recreate these figures without an aerial view, why do it, if not for the purpose of seeing them from above?[63] The encyclopedia states: "Some figures have been measured: the hummingbird is 93 m (305 ft) long, the condor is 134 m (440 ft), the monkey is 93 by 58 m (305 by 190 ft), and the spider is 47 m (154 ft)."[64] Some have theorized these figures are markers for some alien people coming here. Erich von Däniken, in his book *Chariots of the Gods,* tried to make that claim, but it has been refuted.[65] He was right in one way, it was influenced by beings in chariots, but they are demons, not Gods.

Many other places defy explanation on how they were built. One such area is Sacsayhuaman which is also in Peru. It consists of huge stone walls and other structures

---

[63] "Nazca Lines," *Wikipedia*, the free encyclopedia. https://en.wikipedia.org/wiki/Nazca_Lines Accessed 7.6.2021
[64] Ibid.
[65] Ibid.

that defy human ability to construct them. Some of the stones weigh up to 125 tons.[66] That's 250,000 lbs. if my math is right. Moving such stones today would be nearly impossible, using the most advanced equipment. Nearly impossible for humans to do on their own, but Demonic Aliens, easy as pie! Pictures on the page show stones fit together with precision in ways that defy imagination. Using primitive tools you could not do this task. A hammer and a metal chisel? How would you do it? Today we can't do anything similar to it! We use concrete to build buildings and bridges today. The concrete blocks don't fit together with anything like what they did there. So how did they do it? Impossible for humans to do, but Demonic Aliens or giants, no problem! You get the picture here. Man on his own can't do these things, but Demonic Aliens can do them. Not only that, but how did they move the stones from where they were quarried to the wall? After you move it then you have to sit it on top of the previous stone, using only primitive tools. All of these things combined demonstrate that we have been manipulated or at least influenced, if not controlled by a force outside of our realm of humanity. An alien spacecraft with a lifting device of some kind could move these stones, so that is my theory. Demons did it! That's my story, and I am sticking to it!

The Pyramids of Egypt are another area that defies the human ability to construct them. The math is extremely advanced that would allow them to be built that it is difficult to imagine how they were built, apart from an outside influence of Demonic Alien Overlords. Carving each

---

[66] From "Ticket Machu Picchu," "Sacsayhuaman: everything you need to know about the Inca fortress," https://www.ticketmachupicchu.com/fortress-sacsayhuaman/ Accessed 7.7.2021

block used to construct the Great Pyramid and other structures, defy human abilities. One article states:

Did you know that the Great Pyramid is aligned true north? The Great Pyramid is the most accurately aligned structure in existence and faces true north with only 3/60th of a degree of error. The position of the North Pole moves over time and the pyramid was exactly aligned at one time. But not only that, the Great Pyramid of Giza happens to be the center of land mass: The Great Pyramid is located at the center of the land mass of the Earth. The east/west parallel that crosses the most land and the north/south meridian that crosses the most land intersect in two places on the Earth, one in the ocean and the other at the Great Pyramid.[67]

Even today we don't know these things. We couldn't build such a structure using even our most advanced tools. The blocks used to construct it weigh on average over 2.5 tons and there are more than 230,000 of them.[68] How would we move one of these blocks today with our gas-powered heavy equipment? Could we even do it today? How did they carve them out of rock with chisels and hammers? How did they do it? Difficult, if not impossible for man, for Demonic Alien Overlords, no problem!

The objective of these Demonic Alien Overlords is for us to worship them, not God. They have constructed

---

[67] From "Ancient Code," "The Great Pyramid of Egypt a Mathematically Encoded Structure."
https://www.ancient-code.com/the-great-pyramid-of-giza-a-mathematically-encoded-structure/ Accessed 7.7.2021

[68] "Pyramids," "Nova," https://www.pbs.org/wgbh/nova/pyramid/geometry/blocks.html#:~:text=More%20than%202%2C300%2C000%20limestone%20and%20granite%20blocks%20were,block%20is%20about%202.3%20met-
ric%20tons%20%282.5%20tons%29. Accessed 7.7.2021

massive pyramids, huge stone walls, and other structures that are super-human. Even more strange is the fact that some of the ancient Egyptian skulls are also elongated like the ones found in other areas of the world. This could be an indication of the Demonic Aliens also mating with them to produce an alien-human hybrid. Articles indicate that some pharos had elongated skulls. They attribute this to a practice of head binding, of course, but more likely, in some cases, it was produced by a "male" alien mating with a human female to produce the effect, or at least genetic manipulation. To my knowledge, there has been no real DNA analysis of these mummies, but if they did, they might find out the same results as they find for the skulls in Peru, that the father is unknown, and this would prove my theory. In "Psychology Today," it attests to the fact that at least some of the pharos had elongated skulls.[69] Why would they? What is the purpose of doing this practice? Of course, they attribute this to "head binding" not Demonic Aliens breeding with humans, I have written evidence for human-alien hybrids from the Bible, DNA testing of skulls in Peru, and now the UFOs abducting people and taking genetic material. Maybe the real truth is they succeeded in the past with their goal of alien-human hybrids! The pyramids being built would have been an almost impossible human task, but for demonic alien-human-hybrids, not a problem! The Article states:

Given the extensive yet sporadic occurrence of intentional skull modification through head binding, this practice may explain the unusual head shapes of the Pharaoh Akhenaten and his immediate family: his Great Royal

---

[69] Psychology Today," "Strange Head Shapes: Revisiting Nefertiti, Akhenaten and Tut," by Robert D. Martin, Ph.D. https://www.psychologytoday.com/us/blog/how-we-do-it/201907/strange-head-shapes-revisiting-nefertiti-akhenaten-and-tut Accessed 7.7.2021

*Wife Nefertiti, all of their six daughters, and his son Tu-tankhamun. Numerous statues and reliefs indicate conspicuous **head elongation** in all, although this has often been attributed to a new artistic convention.*

On first seeing a side view of Elisabeth's bust of Tutankhamun, I immediately realized that his head must have been tightly bound early in life. Adding to his elongated, flat-topped head shape, a telltale small <u>depression</u> on the crown corresponded to a depression close to the back of the neck. The bust convincingly matches the skull shape reconstructed from a CT-scan. The KV55 skull — whoever its owner may have been — shows a very similar shape, although the head has yet to be reconstructed.[70]

Head elongation seems to be a factor of the rulers of the people whenever there is a huge pyramid, wall, or massive man-made structure that defies human abilities. This is an indication of a possible genetic anomaly at least, if not an actual alien-human hybrid. People of different cultures throughout the world that never had contact with other cultures, use head binding to deform their baby's skull. That is a very difficult conclusion to arrive at when we observe and catalog all of the skulls found. Head binding obviously accounts for some, but not all. Why would they do that, unless they were trying to imitate what the aliens look like? We will never really know why.

[70] "Psychology Today," "Strange Head Shapes: Revisiting Nefertiti, Akhenaten and Tut," by Robert D. Martin, Ph.D. <u>https://www.psychologytoday.com/us/blog/how-we-do-it/201907/strange-head-shapes-revisiting-nefertiti-akhenaten-and-tut</u> Accessed 7.7.2021

Chapter 9
SUPERNATURAL ALIEN ABDUCTIONS

Throughout history, especially recent history, there are increasing incidents of Demonic Aliens abducting people, and the stories they tell are riveting. They never say they are demons, but I am saying it. Even if they are only aliens, their actions are those of a demonic type, so we might as well cut to the chase and call a spade a spade. Anytime someone is kidnapped, taken aboard a spacecraft, has genetic information extracted, sperm or eggs collected, and implants planted on them, it can't be called benevolent, even though the person wasn't killed. This is what we do to the cattle we want to eat! Farmers may pen cattle up, inject them with drugs, milk them, have sperm collected from prize bulls, etc., etc., etc. So, you can see the correlation of what we do with our cattle to what these creatures are doing to us. Call it what you will, I will call it demonic! We are not cattle, we are human, but we are being treated like cattle. And the result is nothing short of the goal of at least domination of our species, but more likely the ultimate goal is, as I have stated before, creating an alien-human-hybrid, which corrupts what God created us to be, and makes us demonic soldiers of Satan! One of the best-documented cases of abduction is the Travis Walton story. Travis Walton wrote a book entitled *The Walton Experience* and later a movie was made entitled *Fire in the Sky* which details the incident.[71] He was missing for over five days and six hours, while working in the Apache-Sitgreaves National

---

[71] From "Travis Walton UFO Incident," *Wikipedia*, the free encyclopedia, https://en.wikipedia.org/wiki/Travis_Walton_UFO_incident Accessed 7.8.2021

Forests near Snowflake, Arizona.[72] His story garnered national attention when as the article states:

According to Walton, on November 5, 1975, he was working with a timber stand improvement crew in the Apache-Sitgreaves National Forest near Snowflake, Arizona. While riding in a truck with six of his coworkers, they encountered a saucer-shaped object hovering over the ground approximately 110 feet away, making a high-pitched buzz. Walton claims that after he left the truck and approached the object, a beam of light suddenly appeared from the craft and knocked him unconscious. The other six men were frightened and supposedly drove away. Walton claimed that he awoke in a hospital-like room, being observed by three short, bald creatures. He claimed that he fought with them until a human wearing a helmet led Walton to another room, where he blacked out as three other humans put a clear plastic mask over his face. Walton has claimed he remembers nothing else until he found himself walking along a highway five days later, with the flying saucer departing above him.[73]

The article continues:

In the days following Walton's UFO claim, The National Enquirer awarded Walton and his co-workers a $5,000 prize for "best UFO case of the year" after they passed polygraph tests administered by the Enquirer and the Aerial Phenomena Research Organization (APRO). Walton, his older brother, and his mother were described by the Navajo County, Arizona sheriff as "longtime students of UFOs". Some UFOlogists believe Walton was abducted by aliens. UFOlogist Jim Ledwith said, "For five days, the authorities thought he'd been

---

[72] Ibid.

[73] Ibid.

murdered by his co-workers, and then he was returned. All of the co-workers who were there, who saw the space-craft, they all took polygraph tests, and they all passed, except for one, and that one was inconclusive."[74]

What were they doing with him for five days? Having a picnic? Trading stories about their planet? I contend they were extracting genetic material or worse to use to produce their human-alien-hybrids! Whatever they did with him, they still kidnapped him. If someone else, did it, it would be at least life in prison! Say it is a hoax if you want, but his co-workers were about to all be charged with murder! Would you risk going to prison for life on a dumb story, and $5,000 divided seven ways? In the movie, when they found him he was naked, in very cold weather at a gas station along a deserted highway at night. Also, at the end of the movie, they all passed the poly-graph test.[75] Later the article states:

Thirty years after the book's release, Walton appeared on the <u>Fox</u> game show *The Moment of Truth* and was asked if he in fact was abducted by a UFO on November 5, 1975, to which he replied, "Yes". The polygraph test determined he was lying.[76]

I say he was not lying because if he were, he went through a huge amount of ridicule for just a little fame. Again, I ask the question, what are these Demonic Aliens doing to people if not "harvesting" us for their own sick use!

There are numerous examples of people being ab-ducted by these creatures. Another example is where Betty and Barney Hill were taken aboard their spacecraft. Wikipedia states:

---

[74] Ibid.
[75] From *Fire in the Sky,* directed by Robert Liberman, 1993
[76] Ibid

They also had missing time, which collaborates their story as the encyclopedia continues:

---

[77] From "Betty and Barney Hill," *Wikipedia,* the free encyclopedia, https://en.wikipedia.org/wiki/Barney_and_Betty_Hill Accessed 7.8.2021

Although the Hills noted that they arrived home later than anticipated, the drive should have taken about four hours (178 miles). They claimed not to have realized that they arrived home seven hours after their departure from Colebrook. When Hohman and Jackson noted this discrepancy to the Hills, the couple had no explanation (a phenomenon ufologists call "missing time"). The Hills claimed to recall almost nothing of the 35 miles of US Route 3 between Lincoln/Indian Head and Ashland. Both claimed to recall an image of a fiery orb sitting on the ground. Betty and Barney reasoned that it must have been the moon, but Hohmann and Jackson informed them that the moon had set earlier in the evening.[78]

Let us see now, they are abducting people, and they are supposed to somehow be benevolent because they didn't kill them. They only used them for whatever they wanted to, and then they let them go. We do this to our farm animals when we need to. Are we farm animals being set up for slaughter? If they are only aliens, they are studying us, finding out our weaknesses, and preparing themselves to take over our world. If, as I show evidence for here, they are the fallen angels (demonic entities) they are developing demon-human-hybrids that will no longer be human, beings that they will be able to rule over and corrupt our genetics! They tried it before, when they mated with human women and made giants that were ruling regular people, The real purpose though was to corrupt the human genetic codes to the Messiah that would come later. That is why the flood happened and only Noah and his family survived. He was evidently the only person left, along with his sons and daughters-in-law that

---

[78] Ibid.

had not been corrupted by the plot to corrupt God's perfect creation!

People don't accept that the Flood actually happened, but I contend it did. Jesus said it did.

> **They did eat, they drank, they married wives, they were given in marriage, until the day that Noah entered into the ark, and the flood came, and destroyed them all.** Luke 17:27

There is also abundant archaeological evidence to demonstrate the Flood actually happened, whether you believe it or not. I will refer you to the excellent work *The Genesis Flood,* by Whitcomb and Morris that proves it. You can also check out my book, *Finding Proof of Jesus,* which details the evidence for what God did to stop the corruption of what he had made by these demons in the past. We know they are obviously still trying to do it again, because one thing you must admire about Satan, is he never gives up!

Things will get worse the closer it gets to the end. They are now trying to make the Covid-19 vaccination mandatory for everyone! The last time I checked it was not going to be <u>required</u>. We are moving into the Hell of our own creation. Maybe it does contain a tracker in it that will help the Demonic Aliens "track" us and control us! Maybe it is the implant of the Mark of the Beast, disguised as a mild-mannered vaccine! I didn't think it was, but if they are going to force us to take it, then that's just what it talks about in the End Times! You won't have a choice, take the mark or be arrested. If you want to work, you have to get it. If you want to buy food, you must take it. If you don't want to be beheaded, you will take it! Then, after you take it, a mark starts to form on your hand or your forehead! You see where I am going with this, right? We are definitely in the Last Days now!

I have been a failure so many times in my life! I only hope I can hold out until the end of my life without folding completely into a subservient slave, doing everything required to survive, not thinking about the consequences. I did what they said to get a driver's license, buy my house, pay my taxes, go along with the laws and try to obey the government that sets the laws. Now I am not so sure I am doing the right thing. Luckily, I am no longer working, or I'm sure they would say get the vaccine or you will be fired. For most of this year in 2021, you had to wear a mask to go into any grocery stores if you wanted to buy food. Now, what if they set up indicators at the entrances of the store, which go off when you enter if you haven't taken the vaccine? They have alarms already when you exit if you didn't buy something. Why not if you haven't taken the vaccine? You will no longer be able to buy or sell without it! They could easily do it. They might just issue people that have the vaccine a card to carry with them they use to swipe to buy food. If you don't have the card, you can't buy any! They are at the very least checking to see who they will have trouble with when the real deal happens with the real implant. Your name will go on the list, and they will arrest you when you refuse to get the next vaccine, then you will know that will be the Mark of the Beast!

I am going to repeat myself again here. Take head of this one scripture in the Bible, which is coming true right before our eyes!

¹⁶ **And he causeth all, both small and great, rich and poor, free and bond, to receive a mark in their right hand, or in their foreheads:**

¹⁷ **And that no man might buy or sell, save he that had the mark, or the name of the beast, or the number of his name.**

¹⁸ **Here is wisdom. Let him that hath understanding count the number of the beast: for it is the number of a man; and his number is Six hundred threescore and six.** Revelation 13: 16 - 18

Never before have they forced everyone to get a vaccine or do anything like they did with the masks. It is showing these Demonic Alien Overlords who they need to "work on." Maybe they will begin capturing those and abducting whoever they are informed on that did not get the vaccine. If you have already received the vaccine, I have no indication it is the implant yet, so you can at least relax for now. But, when the next pandemic happens, and it will, then all bets are off! You will have to get the implant with the vaccine, or else!

Demonic Alien Overlords flying above us in the skies, people on the ground going door to door to find out if you have the vaccine or not, a requirement to wear a mask if you want to buy food, a passport that indicates you have taken the vaccine if you want to fly internationally, at least, and other restrictions. Sounds like a very precarious time, unlike anything we have ever experienced before! It can only get worse. Once they make it mandatory to take the vaccine, then you know it is the end!

I know, I know, I haven't convinced you these aliens are demons, but what else are they? What planet could they be from that we can verify? You also have no proof of that, either. Can you travel to their solar system, and the planet they say they are from, and verify it? Can you trust the news media and the government to be telling us the truth about them? Can you really know anything for sure about them? They will, of course, lie about every-thing to conceal their true objective. I know one thing, since the Bible talks about demons and not aliens, that is what they are! The government has been lying to us from the beginning about these things. Remember for years, they were "swamp gas" or something else. They never told us they were real before. They are lying now in not telling us that the government knows who and what they are. A half-truth is still a total lie! Tell me again how the Roswell Crash was a weather balloon! Tell me again there were no bodies discovered! Tell me again one did not survive! Tell me lies, tell me sweet little lies as the song goes, I will believe you!

But how do they "die" if they are demons, and they live forever? Their bodies die, but they don't ever die spiritually. Just like we don't either, we continue to live on, even after death, either in Heaven or Hell. There, I said it, not politically correct at all, and in some cultures, I could be arrested for saying such a thing! Most people will go to Hell because they won't believe it. The indoc-trination began way back when I was in grade school and even before. They began to teach evolution, not creation. Today they are poisoning young people with all manner of lies about themselves. Your success or failure is deter-mined somehow by your Race, not your character. If you are White, you are evil. They want to ignite a Race war with Critical Race Theory. I say get rid of Race! We are all only one Race, the Human Race! We are all descend-ants of Noah, so there is no distinction in who we are! If they taught that, it would stop all the crap! The Demonic Alien Overlords are surely working overtime to destroy us! They keep us from looking in the Bible for answers, and glue us to YouTube and Twitter! The indoctrination and defeat are nearly complete!

Chapter 10
DEMONIC INFLUENCE TODAY

Today we are powerless to stop all the riots and madness that is going on in our major cities. Riots and chaos everywhere under the guise of peaceful protest. We are accepting the mantras of "Defund the Police," that keep order as the norm. The mayors in these cities allow people to loot and destroy businesses because we somehow "owe" poor people restitution. This is a clear example of the influence of demons on our society! And I contend it is from the Demonic Alien Overlords, that are streaking across our skies that are in control of people. I cited the Scripture that stated it is in "high places" and that could also mean, from above, so that is an indication that demons are inhabiting our skies. So, these alien craft we see are the vehicles they now use to transport themselves since they now have humanoid bodies which they now inhabit. I could be totally wrong, but I'm not! These **aliens are demonic** and nothing else needs to be said. They have yet to reveal themselves completely. They hide in the cover of darkness. They kidnap people out of their homes or when they are driving. They implant devices on people that have a radio frequency they can use to track people. Sounds exactly like what a demon would do!

Now demons control people in ways that make them do all manner of evil. We don't have to look very far at all to see that. We can call it a mental disorder if we want because demons don't exist today since we evolved and are not created, says the scientist, so we can sweep it under the rug, but the things people do are absolute evil. Take for example the case of Jeffery Dahmer. He was an American serial killer sentenced to fifteen life sentences after <u>murdering and eating</u> up to seventeen men and boys.[79] This is absolutely demonic behavior without question! They called him mentally sick, with what is called a schizotypal personality disorder,[80] but we must call what he did demonic. If this behavior is not demonic, then nothing can be categorized as demonic. If this isn't proof that demons exist, you are kidding yourself! The possession of people today is passed off as a mental disorder, but "the truth is out there," as they say! Demonic Alien Overlords traveling is what we call spacecraft, and reigning down over us, causing us to do evil, fit the bill completely as the cause of our sinfulness. It doesn't mean we are not responsible for our actions, but I am just explaining how we might be being manipulated to commit evil acts. We have to find a way to resist the actions we are tempted to do, by these Demonic Aliens hovering over us on high, but most people do not have the strength to resist. Those who are Christians and attend church regularly will have a better chance at resisting, but those who make no effort to resist will fail quickly.

Now you are asking the question, "If it is all demonic control by these Demonic Alien Overlords, how can medication help those who are diagnosed with paranoid schizophrenia, for instance?" Now, I didn't say everything was demonic. There are illnesses that are indeed just physical maladies. There are also demonic influences by Demonic Alien Overlords, which are evident throughout our society now and in the past. We need to recognize the difference when we can and call a spade a spade. You know you are constantly being bombarded with temptations from somewhere. Most of the time for men, it is constant sexual perversion temptations. For women, they

<hr>

[79] From "Jeffrey Dahmer," Wikipedia, the free encyclopedia. https://en.wikipedia.org/wiki/Jeffrey_Dahmer Accessed 7.10.2021
[80] Ibid.

have other temptations, which I don't know what they are, but may also be sexual. For children, they are constantly being tempted to disobey their parents, skip school, or whatever. The temptations fit our age, stage in life, sex, and whatever our propensity is. Some people have trouble with steeling, others not so much. The Demonic Alien Overlords are gearing our temptations in the way they know they will work for each one of us. Tell me again why I am wrong! I know I am right! The evidence is mounting and clearer every day. There are literally millions of examples of demonic manipulation in our lives, resulting in evil, so many that I can't write about them all. But, consider also the case of John Wayne Gacy. He was responsible for the murders of at least thirty-three[81] young men and boys. According to Wikipedia:

"John Wayne Gacy (March 17, 1942 – May 10, 1994) was an American serial killer and sex offender known as the Killer Clown who assaulted and murdered at least 33 young men and boys. Gacy regularly performed at children's hospitals and charitable events as "Pogo the Clown" or "Patches the Clown", personas he had devised. He was also active in his local community as a Democratic Party precinct captain, and building contractor.

According to Gacy, he committed all of his murders inside his ranch house near Norridge, a village in Norwood Park Township, metropolitan Chicago, Illinois. Typically, he would lure a victim to his home, dupe him into donning handcuffs on the pretext of demonstrating a magic trick, then rape and torture his captive before killing him

---

[81] From "John Wayne Gacy," Wikipedia, the free encyclopedia. https://en.wikipedia.org/wiki/John_Wayne_Gacy   Accessed 7.10.2021

by either asphyxiation or strangulation with a garrote. Twenty-six victims were buried in the crawl space of his home, and three others were buried elsewhere on his property. Four were discarded in the Des Plaines River."[82]

I ask you to explain what would cause this evil. Could it be a mental disorder? Not hardly! Look at the word devil and take off the "d" and you get evil. Evil like this is just unexplainable by any medical term. The "devil made me do it," is indeed a truth most educated people are trying to overlook, because if we can show the devil exists, we have to accept the existence of God. When the aliens are disclosed, the educated people will jump at the chance to acknowledge them as aliens, therefore evolution is true, and there is no God. But they miss the real truth, these aliens are demons, not aliens for outer space. They have disguised themselves very well, only I hope you are getting the picture now and coming around to the truth. They are the demons that were once angels in Heaven, which fell and rebelled with Satan, and now are still trying to corrupt God's creation and convince us again, we can also be like God!

I will leave you in this chapter with one more example that will also hit home with demonic influence being the cause of evil, therefore these Demonic Alien Overlords as the cause. I mentioned him before: Ted Bundy. He was evil personified as well. He got his start by viewing pornography, per Wikipedia. Mr. Bundy stated:

"On the afternoon before he was executed, Bundy granted an interview to James Dobson, a psychologist and founder of the Christian evangelical organization Focus on the Family. He used the opportunity to

---

[82] Ibid

make new claims about violence in the media and the pornographic 'roots' of his crimes. 'It happened in stages, gradually', he said. 'My experience with ... pornography that deals on a violent level with sexuality, is once you become addicted to it ... I would keep looking for more potent, more explicit, more graphic kinds of material. Until you reach a point where the pornography only goes so far... where you begin to wonder if maybe actually doing it would give that which is beyond just reading it or looking at it.' Violence in the media', he said, 'particularly sexualized violence', sent boys "down the road to being Ted Bundys." The FBI, he suggested, should stake out adult movie houses and follow patrons as they leave. 'You are going to kill me,' he said, 'and that will protect society from me. But out there are many, many more people who are addicted to pornography, and you are doing nothing about that.'" [83]

People can say what they want about this, but evil is what evil does, and we are all susceptible to fall prey to the evils we allow ourselves to succumb to, which in my assessment are from the demonic manipulation of these Demonic Alien Overlords! If you are viewing pornography, even only as a sometime thing, I will adjure you to stop it, and when you can't, trust that it is part of the influence of these demons that are just doing their "job" to corrupt you; keep you useless from being an instrument for God, if you are a Christian. If you are not a Christian; keep you from accepting Christ, and ultimately take you to Hell with them in the end. Stop doing all the sin you do, which you know you can't, so try again to say it is not

---

[83] From "Ted Bundy," Wikipedia, the free encyclopedia. https://en.wikipedia.org/wiki/Ted_Bundy Accessed 7.10.2021

demonic pressure from you know who, and you will have to admit my theory is sound. Even if it is not from an alien influence, we have to acknowledge it is from the devil, which can masquerade as the Bible says, "an angel of light." What better description for these Demonic Alien Overlords, than as "angels of light?" Nothing else even comes close.

Chapter 11
THEIR GOAL IS TO KILL US!

In the movie *Independence Day* the president is interviewing a captured alien and asks the question:

President: "Can we negotiate a truce? Is there room for co-existence?"

(No answer)

"Can there be peace between us?"

Alien with a tentacle around the scientist Okun: "Peace, no peace."

President: "What do you want us to do?"

Alien with a tentacle around the scientist Okun: "Die."[84]

They wanted us to just die! It is the same with these Demonic Aliens. They, too, want us to just die. The only thing that makes it so they are not able to kill us is that right now they are limited in what they have the authority to do to us, in many cases. In some cases, they may be able to just kill people indiscriminately. Take for example in the Bible when it talks about Job. Satan was withheld from killing him.

> [6] **And the LORD said unto Satan, Behold, he is in thine hand; but save his life.** Job 2:6

So, we may also be under a divine protective order. The abilities of these Demonic Alien Overlords are in effect limited in their scope. I do not have an answer as to why this is not always the case, and why children sometimes die while they are still innocent. Just to be sure, the world is evil, and in such an evil place, the consequences of sin are shown, but the limitations of Satan are also

---

[84]*Independence Day,* co-written and directed by Richard Emmerich. 1996.

shown. Take, for instance, in my own life when walking home from kindergarten, a big kid threw me in front of a truck, and I rolled over the other side and up on the grass before it hit me. The truck just kept going like nothing happened. Satan tried to destroy me, but God limited his power.

I know what you are thinking, that was not Satan, just a kid that was a bully. I say the act was contributed to by the influence of these Demonic Alien Overlords. Sin is not just a thing to be tolerated and overlooked. What if I had died on that day? Would they be able to say it was just an accident? If what we do causes injury to others, and it is a direct result of some sin we are doing, then what is the cause of sin. We react perhaps from an outside influence, not from within ourselves. Our destructive sinful tendencies are possibly being manipulated by Demonic Alien Overlords, hovering above, and directing us to do unspeakable acts. I am going to repeat myself again and cite these Scriptures:

> **And no marvel; for Satan himself is transformed into an <u>an-gel of light.</u>** 2nd Corinthians 11:14
> **Finally, my brethren, be strong in the Lord, and in the power of his might. Put on the whole armour of God, that ye may be able to stand against the wiles of the devil. For we wrestle not against flesh and blood, but against principalities, against powers, against the rulers of the darkness of this world, against spiritual wickedness in <u>high places</u>. ...** Ephesians 6: 10 - 18
> **[7] And there was war in Heaven: Michael and his angels fought against the dragon; and the dragon fought and his angels,**
> **[8] And prevailed not; neither was their place found any more in Heaven.**
> **[9] And the great dragon was cast out, that old serpent, called the Devil, and Satan, which deceiveth the whole world: he was cast out into the earth, and his angels were cast out with him.** Revelation 12: 7 - 9

These "angels of light," hovering above us in "high places" fit the bill exactly when talking about aliens that are now coming down here to oppress us and kill us if they can! I am convinced there must be some kind of electromagnetic wave that constantly shoots down from these crafts, to cause us to sin. We don't see radio waves, we don't see TV waves, or even wireless Internet waves unless we have a receiver. Who knows what other waves are all around us all the time? CB, walkie-talkie, cell phone, and who knows what else? We are obviously the receiver, in our minds, of the sin waves, broadcast continuously from on high by these Demonic Alien Overlords, which are the third of the angels the fell from Heaven, and are hell-bent on taking us all there with them!

Is there another way it happens? I, and I am sure you, have been setting in church or maybe in your office, and all of a sudden, these horrible thoughts just jump into your head. Where did they come from? Embedded in your psyche, or from another external outside source. This is definitely something we need to investigate. Where do the serial killers get their uncontrollable desire to kill? They just do it, because they want to, or is it from an outside source, compelling them; pressuring them to do it. What about crime in general? It may come from peer pressure to commit the crime, or wanting to impress someone. Many times people have to commit murder to become a part of a gang. It may be another totally unrelated reason. Consider the notorious case of David Berkowitz. His outside influence was in this case, a dog!

Berkowitz was arrested on August 10, 1977 and subsequently indicted for eight shootings. He confessed to all of them, and initially claimed to have been obeying the orders of a demon manifested in the form of a dog belonging to his neighbor "Sam". Berkowitz was found mentally competent to stand trial. He pleaded guilty to second-degree murder and was incarcerated in state prison. He subsequently admitted that the dog-and-devil story was a hoax. In the course of further police investigations, Berkowitz was also implicated in many unsolved arsons in the city.[85]

Even though he later recanted the story of a dog, manifested as a demon, I tend to believe what he said at the beginning. However, the influences of the devil will take many forms. Consider this assertion:

---

[85] From "David Berkowitz," Wikipedia, the free encyclopedia. https://en.wikipedia.org/wiki/David_Berkowitz Accessed 7.11.2021

Berkowitz has been incarcerated since his arrest and is serving six consecutive life sentences. During the mid-1990s, he amended his confession to claim that he had been a member of a **<u>violent Satanic cult</u>** that orchestrated the incidents as ritual murder. A few law enforcement authorities have said that his claims might be credible, but he remains the only person ever charged with the shootings. A new investigation of the murders began in 1996, but was suspended indefinitely after inconclusive findings.[86]

He said he was a member of a violent Satanic cult, and that is why he committed the murders. Whatever the reason, we have demons in there somewhere with every explanation. Hovering above us, masquerading as aliens with big black eyes and bulbous heads, or down on the ground, manifesting themselves in demon-possessed dogs, or other people we know, the results are the same; they convince otherwise normal people to do unspeakable things.

SATANIC CULTS?

But, what about this theory there is a violent Satanic cult operating in our midst? It could very well be true! For years it was rumored that Hillary Clinton participated in some type of Satan worship. Again, there were rumors of her trafficking in child molestation. All manner of evil deeds were being attributed to her, and lots of times there seemed to be good evidence for the claims, but later they were always quashed and went nowhere. Even now, the Jeffrey Epstein case of essentially kidnaping and raping underage girls; and the involvement of Bill Clinton and many other powerful people, seems stalled and is going

---

[86] Ibid.

nowhere. They must all be protected on high by Satan himself! They never seem to receive the punishment you or I would receive for the same actions.

Case in point: If you are a member of the Military or ever served, you know how serious it is if you have a security breach. If you were responsible for it, even on a small scale, you are looking at a prison term in Leavenworth! When I was in the Navy, I knew the rules. We had to make the following statement if anyone asked about nuclear weapons on board our ship: "I can neither confirm nor deny the presence of nuclear weapons aboard this US Navy ship, but every US Navy ship is capable of carrying nuclear weapons." I think that is exactly the verbiage I remember. When the ship I was on went into drydock, they even covered the screw, because that was classified. The security was always extremely high aboard all of the ships, for obvious reasons. It is the same with every branch of the service. If you break any of the rules, you are in jeopardy of serious prison time. Why then, was Hillary Clinton not prosecuted for her crimes of emailing classified top-secret documents and information on a non-secure server? I have to say she must be protected by Satan himself on high! Others have been prosecuted for much lower crimes that seem very trivial compared to what she was alleged to have done. Consider the conviction of Machinist's Mate Kristian Saucier. He took photographs of sensitive areas aboard the submarine he was stationed on with his cell phone and for that was sentenced to one year in prison.[87] The article doesn't say if he emailed them anywhere, but just took pictures on a cell phone. Hillary Clinton emailed classified documents,

---

[87] From "Kevin Saucier," Wikipedia, the free encyclopedia. https://en.wikipedia.org/wiki/Kristian_Saucier Accessed 7.10.2021

even top-secret information on a non-secure server, and received no punishment and was not even indicted. Although he was later pardoned by President Donald J. Trump, the sentence really did not fit the crime, when you consider the extent of the damage that Hillary Clinton did to national security.

Chapter 12
ALIENS ARE HERE AMONG US!

After the crash at Roswell, New Mexico, it was rumored there were alien bodies recovered. One of them was rumored to have survived. Now it seems they are collaborating with us to develop our advanced technology that has exploded in recent years. So, they are helping us to develop technology, but what do they want in return? They are not just going to give it away without getting something of high value in return. They are not stupid and are also not that giving. We wouldn't do that, and neither would they. As I have stated many times, they want our genetic material so they can develop human-alien hybrids, which are in reality _demonic-human hybrids_. They say they are from another planet but obviously are not. There is evidence that their bases are right here on Earth. Now, we also have the testimony that the aliens are in fact collaborating with us at the place known as Area 51. In a video on YouTube, former test pilot John Lear states that his father worked with the aliens there and he was responsible for the invention of the first cell phone and the Lear jet.[88] He also stated that Physicist Bob Lazar also worked with the aliens or had at least seen them.[89] John Lear also stated that he and Bob Lazar also watched alien spacecraft take off from an area near Area 51 and that it is happening every day.[90] If this is actually happening, any country would keep this above top-secret,

---

[88] From "Brad Meltzer's Decoded: Proof of UFOs Revealed." Video stop 14:26 – 19:19 and other areas. https://www.youtube.com/watch?v=RizozLy8kbM   Accessed 7.12.2021
[89] Ibid.
[90] Ibid.

to gain an edge over its enemies. I contend it is, but not with alien creatures, but with demons.

There are also other examples in the present day where unknown creatures have been found that appear to be somewhat human. On creature discovered in Mexico is unknown to science at this time. It was found in 2007 and was at first believed to be a type of small monkey that was skinned. After examination, it was determined it has seven human characteristics with similar hands and feet, but it is the size of a rat, which was caught in a trap.[91] Although it also has a tail it was believed to be able to walk on two legs. ***At the end of the video, it was determined that it has "no identifiable genetic material and the results cannot be explained!"***[92] Obviously, it is a human-alien hybrid (demonic-alien) since it is indeterminable as to what the creature is. The fact that it is so small is not of consequence because it could possibly be an adolescent although the video did not make that conclusion either. No identifiable genetic material! I need to repeat that again. It was alive when it was caught in the trap according to witnesses.[93] It was not conclusively determined to be alien but surmised to be some unknown species or creature evolving according to evolution.[94] I come to a different conclusion based on other evidence for previous genetic manipulation by demons recorded in the Bible. Not all the creatures were necessarily giants that were genetically modified by the demons that mated with

---

[91] From *MonsterQuest*: "HUMANOID TERROR FROM THE SKY," (S3, E21) | Full Episode | History https://www.youtube.com/watch?v=Ke07QVP9zlM&t=375s Accessed 7.12.2021

[92] Ibid.

[93] Ibid.

[94] Ibid.

humans, especially during adolescent stages, I would contend. So, we now have a living example of what is happening right now with the aliens corrupting creation.

MORE ALIEN ABDUCTIONS

According to an article in *Scoop World,* as many as 4,432,880 have disappeared worldwide in the last twenty years and that works out to be 607 persons disappearing every single day![95] At least some of these unsolved disappearances are of course explainable when a body is found, and a murderer is tried and convicted. Or maybe the person is just really living in a different city or on an isolated island. Also, some are the result of human trafficking or other nefarious acts. In my mind, all evil that has resulted in a death or disappearance can be attributed to evil demonic influence, but the world refuses to think that way. We attribute the evil to the person who did the act and not some Demonic Alien Overlord, which is where, I contend, the blame should lie. But what about all the cases that are never solved. When nobody is ever found, and no one is convicted of their murder, why not be open to the possibility that a good percentage of these unsolved cases might be attributed to abductions by demons posing as aliens, where they just did not return the person? With all the evidence we have for alien abductions, we should surmise that some people abducted by these Demonic Alien Overlords are just simply not returned! We have no evidence that would indicate otherwise, when people disappear without a trace, and no

---

[95] *Scoop World,* "4,432,880 Missing Persons Vanished in the Past 20 Years," Monday 26, August 2013, 11:20 am
https://www.scoop.co.nz/stories/WO1308/S00441/4432880-missing-persons-vanished-in-past-20-years.htm   Accessed 7.12.2021.

amount of searching results in finding them alive or dead. How would we know if it is true or not? We can only speculate. I, and I'm sure you as well, have seen cases on YouTube where bodies are found that are also grotesquely tortured, and no murderer is ever convicted. These too may be able to be attributed to Demonic Aliens kidnapping and torturing their victim, for whatever their vile purpose might be. I need to drive home my assertion, that whenever a person does evil, it is not entirely on their own, but under the compulsion of demons "forcing them through coercion" to commit the acts. Now I am not making an accusation there is a "demon under every bush," because that is not exactly how it works. Consider that if demons travel in spacecraft type vehicles above us, which are for the most part unseen and unknown, they could indeed be, for lack of a better scientific explanation, "shooting down rays of evil impressions" upon us from on high, to make us do what they want.

It is all over the Internet about people being abducted at night while they are asleep and then returned back to their rooms in an hour or so. Harvard professor John Mack, as previously stated, believes it is actually happening and there are now thousands of cases just in the United States alone. We know about these cases because the people were returned. Sometimes it even involves very young children, with known cases as young as five years old. That is bad enough to hear, but what if many of the children and adults that are abducted are just never returned? I am again going out on a limb and say it doesn't matter if the Demonic Alien Overlords are abducting children directly, or they are possessing men with the desire to do it, the result is the same. The child is still missing! When a young woman or man is abducted by evil men, the devil is indeed the one we should blame for all of this. I am not trying to excuse the actions of men but to explain the reasons behind it. We are all guilty of some type of sin, and the reason is because we listened to Satan at the beginning, hence we are still under that "spell" now. Let's look at how it describes that at the beginning.

> ¹ **Now the serpent was more subtil than any beast of the field which the LORD God had made. And he said unto the woman, Yea, hath God said, Ye shall not eat of every tree of the garden?**
> ² **And the woman said unto the serpent, We may eat of the fruit of the trees of the garden:**
> ³ **But of the fruit of the tree which is in the midst of the garden, God hath said, Ye shall not eat of it, neither shall ye touch it, lest ye die.**
> ⁴ **And the serpent said unto the woman, Ye shall not surely die:**
> ⁵ **For God doth know that in the day ye eat thereof, then your eyes shall be opened, and ye shall be as gods, knowing good and evil.**
> ⁶ **And when the woman saw that the tree was good for food, and that it was pleasant to the eyes, and a tree to be desired to make one wise, she took of the fruit thereof, and did eat, and gave also unto her husband with her; and he did eat.**
> ⁷ **And the eyes of them both were opened, and they knew that they were naked; and they sewed fig leaves together, and made themselves aprons.** Genesis 3: 1- 7

The Bible talks about Satan, in the first instance, disguised as a serpent, convincing the woman to eat the fruit that was forbidden. Notice the serpent was talking. Satan had obviously entered into it somehow or possessed it, for lack of a better term, to make it talk and then deceive the woman. So, right from the beginning, the devil used an animal to disguise himself. Today he uses fallen men to do evil, and has power over them, because of the fact that we fell at the beginning. Demons then, demons now, only this time in the form of what appear to be aliens! Their disguise can be people, animals, things, what appear to be aliens, or it can even be something more canister. You can take it to the bank!

When Travis Walton was abducted, he was gone for five days![96] His friends and co-workers might well have eventually been charged for his murder. If he had never been returned, there would still be suspicion about how they killed him and disposed of the body. In some cases, where nobody is found, and someone is convicted, could it be they actually were abducted by aliens? We have no way of knowing for sure, only that a person has disappeared without a trace. One study suggests that nearly ***four million*** Americans may have been abducted by aliens.[97] The encyclopedia states:

In 1991, Hopkins, Jacobs and sociologist Dr. Ron Westrum commissioned a <u>Roper Poll</u> in order to determine how many Americans might have experienced the abduction phenomenon. Of nearly 6,000 Americans, 119 answered in a way that Hopkins et al. interpreted as supporting their ET interpretation of the abduction phenomenon. Based on this figure, Hopkins estimated that nearly four million Americans might have been abducted by extraterrestrials. The poll results are available at this external link: <u>Abduction by Aliens or Sleep Paralysis</u>.[98]

---

[96] From "Travis Walton UFO Incident," *Wikipedia,* the free encyclopedia, https://en.wikipedia.org/wiki/Travis_Walton_UFO_incident Accessed 7.8.2021

[97] From "Alien Abduction Claimants," *Wikipedia,* the free encyclopedia.
https://en.wikipedia.org/wiki/Alien_abduction_claimants Accessed 7.13.2021

[98] Ibid.

AS MANY AS FOUR MILLION AMERICANS MAY HAVE BEEN ABDUCTED BY ALIENS!

Can someone say that again? If true, we have a real epidemic on our hands worse than anyone thinks. I have no figures about the entire world but based on these figures it would be astronomical. It could even be more because if they are abducted and never returned, the figures go up immensely. They are just a cold case on a detective's desk. Of these abductions, how many have had an implant placed on their body? To my knowledge, we don't have any figures yet, but if over four million Americans have been abducted, they may all have them but are unaware of the presence of the implant. What we forget with every news report is that these aliens are not aliens, but demons. In the past, we used to think people could be possessed by a demon spirit, but now we are far too sophisticated to think that way. If you ask people today if they believe in God or Satan, many will say no. Few people want to be bothered to think they might actually wind up in Hell with the devil and his angels. If God and the devil both don't exist, then aliens do. They are the only hope, many people think for our sick world. When they arrive, they will assume the role of our Savior in the minds of many millennials! That is what makes our world "ripe for the picking" by Demonic Alien Overlords!

Chapter 13
TRUE CASES OF DEMONIC POSSESSION AND OTHER NEFARIOUS THINGS

If Satan came right out and told you he was going to possess you and take you with him to Hell when you die, you wouldn't go along with it. Or if Demonic Alien Overlords showed up disguised as aliens tomorrow and said they are here to rule us, we wouldn't go along with that one either. This is the reason we need guns in every home, as many as we can get, and as powerful as allowed, to fight the aliens when they arrive and try to take over the world. It happened in the movie *Independence Day*.[99] The aliens were massing on the mother ship to go door to door and conquer us for good.[100] We may well be there one day. But we *already* are there! These creatures are already massing on every area of the Earth, preparing for the final signal to strike all at once and once and for all, and we will be defeated without a shot being fired by our military. They won't come out and say they are going to rule us, but the result will be the same, total world domination. We are beginning to get wind of their plans, but most people won't be ready. Already while I write this, liberals are shouting we don't need any more guns! That is because they are secretly in league with the aliens and aren't aware of it. Their minds are "possessed" at least to a degree with the mindset that aliens would be benevolent, and we don't need to be able to defend ourselves from an alien invasion force or from any other invasion, domestic or otherwise. Au contraire, our military would easily be defeated by an alien invasion force, far more

---

[99] *Independence Day*, co-written and directed by Richard Emmerich. 1996.
[100] Ibid.

technologically advanced from us and that would leave us to fight them!

Most people won't like what I just said. They will try to say I am a fringe kook that needs psychiatric help because there are no "aliens," no "demons," and we don't need guns to defend ourselves from them. That is too far out there. Then, "The truth is _not_ out there," and I need help! But many documented cases show that demonic possession is indeed real, and now officials have also stated that UFOs are real. Since UFOs are real, they have occupants piloting them, and **these are the demons!** We know that's what they are, but the government just has not come out and told us that yet. They will when they are ready to be revealed, but they will say they are aliens. The movie _The Exorcist_[101] wasn't just made up in someone's mind but based on true accounts of demonic possession. The experience I cited previously in this book is only one example of a case where there was a clear attempt at demonic possession. There are other similar cases. Another similar case happened to Louisa.

"The night after she left the hospital, Louisa locked the door to her apartment, secured the window with a wooden bar, and went to bed. As Louisa tells it, she awoke in the darkness to the sound of someone's breathing. It seemed close: She could feel the hot exhales on the back of her right ear and her neck. _There's no way anybody could get into the room_, she thought, lying motionless in her sleeping bag. _How is this possible?_

Thoughts of evil spirits rushed to Louisa's mind. Her grandmother, who was both an American Indian and a devout Catholic, had warned her about them. If Louisa

---

[101] _The Exorcist_, 1973 film based on the book by William Peter Blatty.

ever encountered evil spirits, her grandmother had told her, she should do her best to ignore them, because they feed on attention. Louisa tried, but the breathing continued, a heavy, rhythmic rasp. Then, after a minute or so, she felt a hand brush against her collarbone.

At that sensation, which to this day she cannot account for, Louisa leapt out of her sleeping bag and ran to turn on the light. She swears that as soon as she flipped the switch, she heard a pack of stray dogs break out in wild yelps. By dawn Louisa had cleared out, walking several miles to the U.S. Embassy in Kathmandu. She took the next flight back to Orlando."[102]

Surprisingly, this is almost identical to what happened to me in the previous description of when the demons were trying to possess me. Others have had situations similar to this, or family members may request to have the demonic entities exorcised out of their loved ones. The article goes on to report that "The official exorcist for Indianapolis has received 1,700 requests so far in 2018."[103]

There are also plenty of other cases where demons possess people. In one such case in 1921, a lady named Magdalene was diagnosed as demon-possessed.

"The girl loses consciousness, her ego disappears, or rather withdraws to make way for a fresh one. Another mind has now taken possession of this organism, of these sensory organs, of these nerves and muscles, speaks with this throat, thinks with these cerebral nerves, and that in

---

[102] From *The Atlantic*, *"American Exorcism,"* by Mike Mariani, December 2018 issue.
https://www.theatlantic.com/magazine/archive/2018/12/catholic-exorcisms-on-the-rise/573943/ Accessed 7.13.2021
[103] Ibid.

so powerful a manner that the half of the organism is, as it were, paralyzed."[104]

So, the evidence for demons existing is overwhelming, as countless cases exist, even more, terrifying than these, and now since the government has admitted UFOs are real, and since the spacecraft need to have someone or something piloting them, Demonic Alien Overlords are also real. I could list hundreds of cases of demonic possession, but this book is about proof that these aliens in the UFOs are demons. The evidence is presenting itself that aliens and demons are "one in the same." In the movie *Roswell*, there was a bad storm and a loud lightning clap that was used to indicate when the crash at Roswell happened.[105] There is a Scripture that also describes what might have happened.

Jesus is talking:
And he said unto them, I beheld Satan as *lightning fall* from Heaven. Luke 10:18

Obviously, Jesus is not talking about the Roswell Crash, but this passage indicates that Satan crashes down from Heaven (above). These alien creatures travel around in the Heavens above us and if so they would obviously have some sort of craft to travel in. Let's revisit the fall of Satan in the scripture to determine what demons are. Demons are angels that rebelled against God at the beginning. They were the servant of the Most High God, and they rebelled. How this could happen, I don't have

---

[104] From *The Atlantic*, *"American Exorcism,"* by Mike Mariani, December 2018 issue.
https://www.theatlantic.com/magazine/archive/2018/12/catholic-exorcisms-on-the-rise/573943/ Accessed 7.13.2021
[105] *Roswell*, 1994 television film produced by Paul Davids.

an answer, but go with it and try to put yourself in the place of angels who serve God. One day the main angel, Satan, has gathered together a group of other angels to rebel against God. Think of it like a mutiny on a ship. Some of the crew are with the mutineers, and some are not. They kept their mutiny plan secret from the captain and those who would not go along with the mutiny. How they did that when God knows everything is beyond my understanding, but they did not succeed, so God knew about it.

> **⁷And there was war in heaven: Michael and his angels fought against the dragon; and the dragon fought and his angels,**
> **⁸And prevailed not; neither was their place found any more in heaven.**
> **⁹And the great dragon was cast out, that old serpent, called the Devil, and Satan, which deceiveth the whole world: he was cast out into the earth, and his angels were cast out with him.** Revelation 12: 7 - 9

Satan wants to be God. The Bible describes it like this.

> **¹²How art thou fallen from heaven, O Lucifer, son of the morning! how art thou cut down to the ground, which didst weaken the nations!**
> **¹³For thou hast said in thine heart, I will ascend into heaven, I will exalt my throne above the stars of God: I will sit also upon the mount of the congregation, in the sides of the north:**
> **¹⁴I will ascend above the heights of the clouds; I will be like the most High.**
> **¹⁵Yet thou shalt be brought down to Hell, to the sides of the pit.** Isiah 14: 12 - 15

Satan failed at being God. Now his goal is to corrupt God's creation. Now he wants to keep trying to do what he failed at in the beginning. Since he will ultimately be going to Hell in the end, the only thing he can do now is try to take as many people with him as he can. He can also get them to worship him, by keeping them from believing in God and lying about creation. If the masses believe we evolved, then God is out of the picture. With God out, Satan moves in by getting us to believe that aliens somehow "seeded" the Earth and evolution took over. These alien beings move in to claim the place of God! That is why we have to call them what they really are, not aliens but demons, hell-bent on taking the place of God. In the past, they used demon possession to take over bodies. Now they have developed the humanoid bodies we see all over the Internet; small frail looking creatures with bulbous heads and big black eyes; skinny little arms with only four fingers and tiny little legs. They travel about in their spacecraft, for the most part, unseen and unheard, but at times they must have to expose themselves to do certain tasks, so that would be the only reason we are seeing them.

I surmise, for the most part, they are like the Klingons on Star Trek where their spacecraft have a cloaking device, except when they want to fire on another spaceship, they must expose themselves. Likely, when these Demonic Alien Overlords want to abduct people, or do other things, they need to expose themselves to do certain tasks. You will never see them at all unless they need to do something they can't do while "cloaked." Almost all the time they are seen is also at night. Although they have been seen during the day, often it is because they may have some issue with their spacecraft. Sightings in broad daylight are rare but becoming more prevalent now. That would be an indication that their activity is also increasing because the time is becoming much shorter with every passing day before the Rapture occurs, which will usher in their full disclosure. Now, in order to avoid worldwide panic, they are releasing information about them in stages. This year, 2021, it was just acknowledged that UFOs are real. Next year something new, until finally everything will be disclosed after the Rapture of the church has occurred.

Witches, ghosts, goblins, ghouls, and every other vile thing can be attributed to demons; but I am going out on a limb here and attributing UFOs to demons as well. If there were truly aliens inhabiting other worlds that would arrive on Earth someday to "save" us, God would have given us at least some hint in the Bible, I argue. Since Jesus is the only true Savior, there are no aliens that will show up to save us from ourselves, but there are demons that will try to fit that bill. When they say they are from another planet in another solar system, we will not be able to prove otherwise, so the lie will prevail, and everyone in the media will accept them as what they say they are, and then the world will follow them. Consider if it happened today, **ALIENS LAND AT THE WHITE HOUSE!** The papers would explode in articles about them! It would remain the top story for months or even a year. Laws would be passed to protect them and compel us to follow them, and if we do not we would be arrested and ultimately executed. Call them demons then and you are a heretic! They are the new god, and you are required to worship them. I for one will not worship them but will continue to scream they are demons! The whole world will "wonder after the Beast," just like the Bible says it will!

> **And I saw one of his heads as it were wounded to death; and his deadly wound was healed: and _all the world wondered after the beast._** Revelation 13:3

I know I am repeating myself over and over, but unless I keep harping on it, the fix will be in. I may already be too late, although I am trying to finish this book as quickly as I can. The Rapture could occur any day now and then it will be too late. I have to warn you now before the end! Whatever they try to say about the UFOs or UAPs or whatever they want to call them, just keep repeating over and over to yourself:

"THEY ARE DEMONS AND NOT ALIENS!" "THEY ARE DEMONS AND NOT ALIENS!" "THEY ARE DEMONS AND NOT ALIENS!"

Then when they insist you get the implant, resist with everything you can! Run away to the mountains and hide; they may still find you, but you can hold out for a little while. Heed this dire warning:

> **14 But when ye shall see the abomination of desolation, spoken of by Daniel the prophet, standing where it ought not, (let him that readeth understand,) then let them that be in Judaea flee to the mountains:**
>
> **15 And let him that is on the housetop not go down into the house, neither enter therein, to take any thing out of his house:**
>
> **16 And let him that is in the field not turn back again for to take up his garment.**
>
> **17 But woe to them that are with child, and to them that give suck in those days!**
>
> **18 And pray ye that your flight be not in the winter.**
>
> **19 For in those days shall be affliction, such as was not from the beginning of the creation which God created unto this time, neither shall be.**
>
> **20 And except that the Lord had shortened those days, no flesh should be saved: but for the elect's sake, whom he hath cho-**

The Demonic Alien Overlords will arrive on the scene, claim to be God, and demand you get the implant. Even today as I write this in July 2021, they are talking about going door to door to find out who hasn't gotten the *implant,* I mean vaccination for Covid-19 yet. I thought HIPA laws kept us safe from having to reveal information about our medical status. Will they arrive at your door will they also have someone there with a needle to give it to you? Right now, I don't think this vaccination is the implant yet. I am not telling you to not take the vaccination. But, this is just finding out who will actually resist taking the implant when the implant becomes mandatory. How many people will they have to round up them and take to the FEMA camps? Don't answer the door! It's nobody's business but your own. The government does not need to know if you took the vaccination or not. The next time though, you will have no choice. You might not at this time if you want to work. Many employers are already requiring you to get the vaccination or be fired. See how it works, take the implant, or you can't work, you can't buy food and you can't sell anything. We are arriving at this place at breakneck speed! And you still think the Bible is a myth! I have to repeat it again!

> **16 And he causeth all, both small and great, rich and poor, free and bond, to receive a mark in their right hand, or in their foreheads:**
> **17 And that no man might buy or sell, save he that had the mark, or the name of the beast, or the number of his name.**
> **18 Here is wisdom. Let him that hath understanding count the number of the beast: for it is the number of a man; and his number is Six hundred threescore and six.** Revelation 13: 16 - 18

And who again is the "Beast?" Satan, the ruler of the Demonic Alien Overlords fits the bill to a tee! He may in fact be an alien-humanoid-hybrid, rather than just a regular old alien creature. I think that will fit better in the

Scripture than some alien, but people would worship either one. Today, as I stated previously, the United States is working on hybrid humans for the military. DARPA (Defense Advanced Research Projects Administration) is doing just that. If they don't other countries will, so they figure they have to keep ahead of the game. Others have also said this. Author Tom Horn on the television program SkyWatchTV has alluded to this, although I don't have a direct quote at this time. Since the research going on at Area 51 is centering on alien-humanoid-hybrids, it has long been rumored, the beast will most likely come out of there, I believe.

Chapter 14
WHY TRUE ALIENS ARE MOST LIKELY A MYTH

We all love the *Star Trek* television program and movies, or *Star Wars* but these are only stories cooked up by people. We really don't have any concrete evidence for real aliens yet. The idea that because the Universe seems so vast, there must be millions of other intelligent life forms out there is really just speculation. I contend that regardless of how vast the Universe is, we are likely the only intelligent life forms there are, and the Earth is the only planet that supports other forms of life. I obviously can't prove a negative, but what I can do is show why it is most likely that true aliens do not exist. We always hear repeated that if we are the only intelligent life, it must be a great waste of space, but that again is mere speculation. I submit that, in order for the Earth to exist as it does, the Universe has to be as vast as it is. I am not the only one saying this, but others do also. I just remember someone else making that claim, although I can't find a quote now. Think about if for a moment, our planet is perfectly suited to support life as we know it. Mars, on the other hand, is highly inhospitable, as are all the other planets in our known Solar System. If life existed anywhere else in our Solar System, Mars should be a prime candidate, but nothing has been found as yet with our explorations and probes. I contend, there is no life there because it is not suitable for life as we know it. Although they keep searching for planets they think could possibly support life, they again are speculating. Even if evidence of intelligent life is discovered on Mars, like some so-called alien artifacts or creatures like these UFO aliens that we have here, they are not true aliens, but demons. I keep repeating that because I need to keep reminding everyone of this one fact. Aliens (what people want to call aliens) are demons and not aliens. That is why the Bible leaves out any reference to aliens from another planet. God would not have left that one important fact out if there were aliens from another planet, because he doesn't deal in half-truths. We also would not be prepared for a time when aliens disclose themselves, and many would be unable to cope with their arrival. So, God would not leave us hanging on such an important issue but would have at least left some clues. He did mention dinosaurs, for example, as it references them several times in the Bible, notably in the book of Job chapters 40 and 41.

> [15] Behold now behemoth, which I made with thee; he eateth grass as an ox.
> [16] Lo now, his strength is in his loins, and his force is in the navel of his belly.
> [17] He moveth his tail like a cedar: the sinews of his stones are wrapped together.
> [18] His bones are as strong pieces of brass; his bones are like bars of iron.
> [19] He is the chief of the ways of God: he that made him can make his sword to approach unto him.
> [20] Surely the mountains bring him forth food, where all the beasts of the field play.
> [21] He lieth under the shady trees, in the covert of the reed, and fens.
> [22] The shady trees cover him with their shadow; the willows of the brook compass him about.
> [23] Behold, he drinketh up a river, and hasteth not: he trusteth that he can draw up Jordan into his mouth.
> [24] He taketh it with his eyes: his nose pierceth through snares.
> Job 40: 15 - 24

Sounds like a dinosaur to me. Other areas in this same book talk about what sounds like a sea monster.

<blockquote>
¹ Canst thou draw out leviathan with an hook? or his tongue with a cord which thou lettest down?
² Canst thou put an hook into his nose? or bore his jaw through with a thorn?
³ Will he make many supplications unto thee? will he speak soft words unto thee?
⁴ Will he make a covenant with thee? wilt thou take him for a servant for ever?
⁵ Wilt thou play with him as with a bird? or wilt thou bind him for thy maidens?
⁶ Shall the companions make a banquet of him? shall they part him among the merchants?
⁷ Canst thou fill his skin with barbed irons? or his head with fish spears?
⁸ Lay thine hand upon him, remember the battle, do no more.
⁹ Behold, the hope of him is in vain: shall not one be cast down even at the sight of him?
¹⁰ None is so fierce that dare stir him up: who then is able to stand before me? Job 41: 1 - 10
</blockquote>

So, if God mentioned dinosaurs, something that we don't have today, normally to battle with, he would have also mentioned aliens if they were truly going to be a factor in our lives. He purposely did not mention them because they don't exist, but he mentioned demons, which do exist. That is a very simple explanation as to what these UFO beings are, not aliens, but demons.

Some major scientists want to deny that God is responsible for the creation of the Universe and ultimately us, not necessarily because there is evidence to the contrary, but because if they acknowledge God created the Universe and everything there is, they have to change their behavior. Not only so, but I believe they are trapped in a belief system by Demonic Alien Overlords bombarding them with "sin waves" from above, for the lack of a better term. Satan wants to control the flow of information we receive as well. He wants to keep men from hearing the Gospel that would get them saved, or if they do after it is entirely too late for them to change their belief system. There is virtually overwhelming evidence for creation by God, rather than evolution, but few will acknowledge it. Take a simple mathematical proof that shows the impossibility of an evolutionary construct. We have to start with nothing when we look at the Universe. If we have something then that denotes that something is responsible for the "creation" of it. I give you this most simple equation:

$$0 + 0 = 0!$$

We start with nothing at the beginning, and since nothing can be added, we end up with nothing still at the end. I will not concede this point. Nothing is available to be used to produce evolution. If something is there, then God is responsible for it! You don't get to have matter in any form at all, no speck of dust, no gas particle, no matter compressed down into the head of a pin, nothing. Therefore, you have nothing because you started with nothing. Further, Dr. Kent Hovand, a well-known creationist, has pointed out that time, space, and matter have to come into existence at the same time, or nothing can exist.[106] I am not quoting him exactly, but basically, it is that space, matter and time have to all come into existence all at once or they cannot exist, because if you have matter but no space where would you put it? If you have space but no time, when would you put it?[107] All three

---

[106] Video "Where did God come from?" YouTube
https://www.youtube.com/watch?v=w6AHcv19NIc&ab_channel=ghitamoldovan Accessed 7.15.2021
[107] Ibid.

have to come into existence at the same instance and the Bible describes this creation in the first verse of Genesis:

> [1] **In the beginning God created the Heaven and the Earth.**
> Genesis 1:1

This verse shows time; *in the beginning*; space, *Heaven*; and Earth, *matter*. All are created at the same instant. Additionally, the force (God) that would create such things as space-time and matter has to be outside of these things, and unaffected by them, because if it (God) is affected by it, it (God) could not create space, time, or matter.[108] Additionally, he and other theologians have pointed out that morality can't exist without God. Everything would be relative, and nothing would be wrong or right without God, because there is no standard above ourselves.

Thus enter these so-called aliens who arrive on the scene and try to tell us what is wrong or right. They will try to tell us it is ok to get the implant because once you get it, you will be healed of all diseases. Never mind that you will be their obedient slave. Since evolution is true, they have evolved and there is no God! But since they would have to have evolved, once we show that evolution is false, they too are false! The idea that evolution can occur from nothing is absurd. Scientists have also come up with what is called the Steady-state Model[109] that matter has always existed, but then matter would be God. Per Wikipedia:

In cosmology, the steady-state model is an alternative to the Big Bang theory of evolution of the universe. In the

---

[108] Ibid

[109] "Steady-state Model," *Wikipedia*, the free encyclopedia. https://en.wikipedia.org/wiki/Steady-state_model   Accessed 7.15.2021

steady-state model, the density of matter in the expanding universe remains unchanged due to a continuous creation of matter, thus adhering to the perfect cosmological principle, a principle that asserts that the observable universe is practically the same at any time and any place.[110]

If matter has always existed, then it would be God because it never had a beginning or an end which is indeed a major attribute of God, by definition. If you have more questions about why evolution is false and the Bible is true, you could check out the excellent work, *The Genesis Flood* by Whitcomb and Morris, or my book *Finding Proof of Jesus*. I could cite hundreds of reasons why evolution is false, but this book is not about disproving evolution or proving creation but demonstrating why the aliens we think are arriving in UFOs are really demons here from the beginning of creation and not aliens. I am demonstrating why they are demonic, thus far, for the following reasons:

True Aliens *are not* mentioned in the Bible, therefore do not likely exist.

We could not have evolved, therefore, neither could aliens.

Demons *are* mentioned in the Bible; therefore, any discovered so-called *aliens* are demons.

The lie they will state is that they "seeded" our planet, thus taking credit for God's creation.

The bases they have are here on Earth and not on some other planet, but they will lie and say they are from another planet. Government officials have stated there is no indication UFOs or UAPs are from another planet.

They are abducting people, doing medical experiments on us, collecting genetic material, and all manner

---

[110] Ibid.

of nefarious acts, even collecting sexual materials (sperm and eggs) therefore they are not benevolent aliens but are shown to be demons.

This collecting of genetic material would be for the purpose of making hybrid-alien creatures which is a continuation of what was done before in Genesis when the demons "mated" with human females and produced giants.

Angels are a little higher than us, thus fallen angels (demons) would be more technologically advanced than we are, as these creatures are.

When they are disclosed, people will worship them and forget about God, like the Bible says, the whole world will wonder after the beast.

They are implanting people with devices, which falls into place with the Bible where everyone will be required to get the mark to buy or sell, which the implant could in fact be a type of "mark."

Demonic entities control people and Demonic Alien Overlords flying above us would be in a position to do just that.

Other reasons demonstrating a young Universe.

They are sexually depraved.

SHORT PERIOD COMETS

Further, the Universe can be shown to be young in many ways, apart from scientists asserting huge ages for the Universe, evidence can be presented to show a young Universe, and therefore a young Earth. If the Universe is young and also the Earth, then evolution fails absolutely. Creation works whether the Universe is young or old. Some major factors that show the Universe is young are the short-period comets.

It is a fact that comets break up at a very rapid rate due to gravitational attractions, the solar wind, and external explosions. Given that comets were supposedly formed at the same time as the rest of the universe, the maximum age of a short-period comet is about 10,000 years, according to Dr. Harold Slusher and other astronomers.[111] *"If you determine, therefore, the age of a short period comet, you automatically have the age of our solar system. If we take into account the maximum lifespan of a short-period comet, the solar system could be no more than about 10,000 years old."[112]*

THE AGE OF THE MOON

There is not only this factor that shows a young Universe, but also the age of the Moon. If you assume the Moon was formed at the same time as the rest of the Universe, it should have an exceedingly old age. It can be shown to be exceedingly young.

It has been reported in *Scientific American* by Hans Peterson that some 14 million tons of dust fall to the Earth's surface each year from meteorites that burn up as they enter the Earth's atmosphere. This same rate or greater can also be calculated for the moon.[113]

"In fact when astronauts first landed on the moon, many Evolutionists expected a very thick layer of loose dust. Depth estimates ranged from 50 to 180 feet- and even much higher. Sinking deep into this dust was at least one astronaut's greatest fear about landing on the Moon.

---

[111] *Finding Proof of Jesus*, James L. Kearns, WestBow Press, 2016. P. 110

[112] Edward F. Blick, *Correlation of the Bible and Science*, Southwest Radio Church Edition, 1976, pp.30 - 31.

[113] *Finding Proof of Jesus*, James L. Kearns, WestBow Press, 2016. P. 111

Thus, large saucer-shaped feet were provided for the Lunar Lander."[114]

What was actually found is illustrated in the image of the footprint on the moon: a very small accumulation of dust. This is an indication of an exceedingly young age for the moon and therefore also for the Earth.[115]

THE INCREDIBLE SHRINKING SUN

The sun is shrinking in size. Given that it is assumed that it came into being with the Universe, it poses a problem.

It is also known that the sun is shrinking at a rate of about 5 feet per hour, as observed by over 100 different observers at the Royal Greenwich Observatory and the U.S. Naval Observatory. As far as we know, this rate has been constant since the beginning of the sun's formation. Given the rate at which the sun is shrinking, it would have been twice its known radius only 100,000 years ago. The conclusion that can be drawn from this is that one million years ago or less, no life could have existed on the planet Earth. As early as 210 million years ago, the surface of the sun would have been touching the surface of the Earth.[116]

There are other factors we can use to show a young Universe. Even though we have tremendous distances to travel to reach the outer reaches of the Universe, it can be shown to be young by a theory developed called the Anisotropic Synchrony Convention. It solves the problem of

---

[114] *The Illustrated Origins Answer Book*, Fifth Edition, Eden Communications, Paul S. Taylor, 1995, p. 17.

[115] *Finding Proof of Jesus*, James L. Kearns, WestBow Press, 2016. P. 114

[116] The Illustrated Origins Answer Book, Fifth Edition, Eden Communications, Paul S. Taylor, 1995, p. 16.

the huge distances regarding the age of the Universe because the speed of light may not be constant. A definition is found for this theory at *Creation Wiki* which describes it as:

The Anisotropic Synchrony Convention presumes that the human observer is in a unique location. Light arriving from a distance source travels at a different speed than light traveling from a local source. Likewise the light reflecting off a surface may travel at a different speed than light arriving from an emitting source. This convention was the universal convention of observation prior to the 1600s. It basically means that to human perception, things that are observed are happening now, not at the end of light-speed propagation delay as with the Einstein convention. Light from the Sun arrives at the same time it is emitted. Light from distant stars arrives at the same time it is emitted.[117]

The conclusion which can be drawn from this is, therefore:

Basically, this shows that now the distances are not a problem regarding the young age for the Earth as stated in the Bible—and this even works with the special relativity law of Einstein![118]

Since evidence exists for a young Universe and therefore a young Earth, evolution fails for us as well as for any so-called aliens that are supposed to be out there. I could go on and on as to why the Universe can be shown to be young, but again this is about Demonic Alien

---

[117] *Creation Wiki* "Anisotropic Synchrony Convention," https://creationwiki.org/Anisotropic_synchrony_convention Accessed 7.16.2021.

[118] *Finding Proof of Jesus*, James L. Kearns, WestBow Press, 2016. P. 50

Overlords, so this work is not about evolution, but about aliens that are in fact not aliens, but just demons. Evolution can't work if we don't have huge ages for it. Creation, however, works both with young ages and old ages, and therefore creation is more viable. Without matter, space, and time coming into existence all at once, from an outside influence unaffected by these factors which would be God, the Universe does not exist. Thus, we don't get aliens evolving to create us, but according to what we know from the Bible we have demons, and that is as I stated what these creatures flying around in UFOs are.

Jesus made _us_ the crown jewel of creation in his universe. That is what the Bible says. If aliens somehow exist, they would also be sinful and in need of a savior. Jesus would also have to die for them. If dozens or hundreds or even millions of different alien creatures exist, Jesus would have to continuously die for them as well, as he did for us. Therefore, aliens were not created, but angels that are his servants were. Obviously what these creatures are are demonic angels that fell from their estate and became the demons that plague us now. Aliens, like man, did not evolve but would have had to have been created by God for them to exist. Therefore true aliens most likely do not exist, but we know that God created angels, and these creatures are the demons that rebelled against God.

The only scenario that includes aliens is an evolutionary construct, without God. If God does not exist and evolution is true, then aliens might have somehow evolved. We are hopelessly lost if there is no God, and this life is all there is. I know that God exists, therefore, aliens do not exist and once they are revealed as aliens they will be treated as a type of "god," just what Satan wants- Satan wants to be God - like it says here:

> **For thou hast said in thine heart, I will ascend into Heaven, I will exalt my throne above the stars of God: I will sit also upon the mount of the congregation, in the sides of the north:**
> **I will ascend above the heights of the clouds; I will be like the most High.** Isaiah 14: 13 - 14

So, my conclusion here is that at the end of this age is these creatures will present themselves as aliens from some faraway planet here to help us, but will in fact be fallen angels (demons) and will be "worshiped" by the masses as a god.

Chapter 15
BARNEY AND BETTY HILL OTHERS

There continue to be alien abductions in the United States and worldwide. The first widely reported incident happened between September 19th and 20th of 1961.[119]

The incident came to be called the "Hill Abduction" and the "Zeta Reticuli Incident" because the star map shown to Betty Hill could possibly be the Zeta Reticuli system according to some researchers. Their story was adapted into the best-selling 1966 book *The Interrupted Journey* and the 1975 television film *The UFO Incident*. In September 2016, plans were announced to make a feature film based on the events, with an unknown release date.[120]

There have been many television programs and even a possible movie about the event. After the incident happened the two began to have dreams and recalled the events that happened that night. This account shows what is alleged to have happened.

Approximately one mile south of Indian Head, they said, the object rapidly descended toward their vehicle, causing Barney to stop in the middle of the highway. The huge, silent craft hovered approximately 80–100 feet (24–30 m) above the Hills' 1957 Chevrolet Bel Air and filled the entire field of view in the windshield. It reminded Barney of a huge pancake. Carrying his pistol in his pocket, he stepped away from the vehicle and moved closer to the object. Using the binoculars, Barney claimed to have seen about 8 to 11 humanoid figures, who were peering out of the craft's windows, seeming to

---

[119] "Barney and Betty Hill," *Wikipedia,*
https://en.wikipedia.org/wiki/Barney_and_Betty_Hill   Accessed 7.16.2021
[120] Ibid

look at him. In unison, all but one figure moved to what appeared to be a panel on the rear wall of the hallway that encircled the front portion of the craft. The one remaining figure continued to look at Barney and communicated a message telling him to "stay where you are and keep looking." Barney had a recollection of observing the humanoid forms wearing glossy black uniforms and black caps. Red lights on what appeared to be bat-wing fins began to telescope out of the sides of the craft, and a long structure descended from the bottom of the craft. The silent craft approached to what Barney estimated was within 50–80 feet (15–24 m) overhead and 300 feet (91 m) away from him. On October 21, 1961, Barney reported to National Investigations Committee On Aerial Phenomena (NICAP) investigator Walter Webb that the "beings were somehow not human."[121]

They both later reported being actually abducted by these beings. They both have undergone hypnosis to discover what happened that night and during one session Betty sketched out what was called a "star map" which was later discovered to be of what is thought to be Zeta Reticuli, a particular star formation, which she would have had no knowledge of.[122]

Simon gave Betty the post-hypnotic suggestion that she could sketch a copy of the "star map" that she later described as a three-dimensional projection similar to a hologram. Though the map she saw had many stars, she drew only those that stood out in her memory. Her map consisted of twelve prominent stars connected by lines and three lesser ones that formed a distinctive triangle… She said she was told the stars connected by solid lines

---

[121] Ibid.
[122] Ibid.

formed "trade routes," whereas dashed lines were to less-traveled stars.[123]

Because she was able to draw a map of what later was interpreted to be a star formation in Zeta Reticuli, researchers who believe the accounts assume the abduction was by aliens that travel in spaceships like on *Star Trek*. I believe they were abducted and held hostage for about three or more hours during this night. I just do not believe the abductors are actually aliens, but demons. Ask yourself, "Why would aliens need to abduct people?" Let's see, extremely technologically advanced aliens traveled all the way from the Zeta Reticuli star system to "little old Earth," however many light years that is from Earth, and now need to abduct people and do experiments on them, and who knows what other things were done that are blocked out! Bull Crap! Advanced benevolent aliens with our good in mind would not clandestinely abduct people, but demons would for their purposes of developing demonic human hybrids.

The fact that she was able to draw a picture of a map from a particular star formation, does not prove these creatures are aliens. Demons (fallen angels) are more advanced than we are and would have knowledge of astronomical formations. In perpetrating their "Great Lie," they will leave clues around for us to find, like a mastermind covering up a crime, or someone trying to finger someone else as the guilty party. They lie and say they have also traveled to the Earth, but for what purpose. In the movie *Independence Day*, the aliens planned to kill all of us, exploit all of the natural resources of the Earth, and then move on to the next planet. What these Demonic Alien Overlords want to do is a re-run of the same old

---

[123] Ibid.

movie, only the twist is, they are not aliens, but the demons of Earth!

Even if these were to turn out to be just regular aliens, they are showing their true colors by operating out of <u>Satan's Playbook</u>. The government lies for them about whether they exist or not, they abduct people, collect genetic material, implant things on people, and all manner of evil. A real alien would act differently, we hope, or we will be destroyed once they reach us. Unlike *The War of the Worlds,* or *Independence Day,* we wouldn't win! They, if they can travel the great distances would have technology so far above us, they would wipe us out all at once with the *Star Wars* "Death Star" or enslave us for sure! What assurance do we have those aliens would be nice? We have none, except to hope they are if aliens are real. We do know that demons are evil, and these creatures display absolute evil in what they are doing. To use the old phrase, "If it quacks like a duck.... It must be a duck!

THE CALVIN PARKER AND CHARLES HICKSON INCIDENT

The stories start out almost the same. People see a light of some kind in the sky and when they investigate what it is, they are abducted. *The Mississippi Clarion Ledger* reports the incident like this: *Parker said he noticed blue light reflecting off the water and his initial thought was law enforcement officers had arrived to tell the two fishermen they needed to leave the property. However, when Parker looked up, he realized the light was coming from a craft like nothing he'd ever seen.*

"A big light came out of the clouds," Parker said. "It was a blinding light.

"It was hard to tell with the lights so bright, but it looked like it was shaped like a football. I would say, just

*estimating, (it was) about 80-foot. (It made) very little
sound. It was just a hissing noise."*

*Then the situation became more surreal. Parker said
three legless creatures floated from the craft. One had no
neck with gray wrinkled skin. Another had a neck and
appeared more feminine. Parker described their hands
as being shaped like mittens or crab claws.*

*When one of the creatures put one of its claws around
his arm, Parker said he was terrified, but then another
feeling came over his body.*

*"I think they injected us with something to calm us
down," Parker said. "I was kind of numb and went along
with the program."*
*Parker said the creatures floated he and Hickson into the
craft and performed physical examinations on the two.
Then they were taken back to the bank of the river.*[124]

Notice in this case also, how they saw a light in the
sky, they were abducted, this time possibly by robotic or
hybrid creatures. They had medical experiments performed on them and then they were released. The reason
for the medical experiments is most assuredly to collect
genetic material for developing alien-human-hybrid creatures. This goes back to Genesis chapter six in the Bible.
I will rewrite the Scripture reference here:

> [4] **There were giants in the earth in those days; and also after
> that, when the sons of God came in unto the daughters of
> men, and they bare children to them, the same became
> mighty men which were of old, men of renown.** Genesis 6:4

---

[124] *Mississippi Clarion Ledger*, "Alien abduction: 45 years after alleged UFO encounter, Mississippi man breaks his silence," by Brian
Broom, 8.13. 2018. https://www.clarionledger.com/story/ magnolia/
2018/08/13/alien-abduction-45-years-after-ufo-encounter-missis-
sippi-man-breaks-his-silence/922706002/ Accessed 7.17.2021

Notice the demonic entities mated with human women and giants were born out of that union. In this case, obviously, they are collecting human genetic material *to do the same thing again!* In this case, though, they must be making the hybrids look more human, so they can fool us better. If huge grotesque giants somehow just appeared, we would be alerted to what they are doing. The ultimate human-alien-hybrid would be the "person" who Satan himself inhabits. Notice what it says about him in this Scripture:

> [11] **And I beheld another beast coming up out of the earth; and he had two horns like a lamb, and he spake as a dragon.**
> [12] **And he exerciseth all the power of the first beast before him, and causeth the earth and them which dwell therein to worship the first beast, whose deadly wound was healed.**
> [13] **And he doeth great wonders, so that he maketh fire come down from Heaven on the earth in the sight of men,**
> [14] **And deceiveth them that dwell on the earth by the means of those miracles which he had power to do in the sight of the beast; saying to them that dwell on the earth, that they should make an image to the beast, which had the wound by a sword, and did live.**
> [15] **And he had power to give life unto the image of the beast, that the image of the beast should both speak, and cause that as many as would not worship the image of the beast should be killed.**
> [16] **And he causeth all, both small and great, rich and poor, free and bond, to receive a mark in their right hand, or in their foreheads:**
> [17] **And that no man might buy or sell, save he that had the mark, or the name of the beast, or the number of his name.**
> [18] **Here is wisdom. Let him that hath understanding count the number of the beast: for it is the number of a man; and his number is Six hundred threescore and six.** Revelation 13:11 - 18

Wouldn't a human-alien-hybrid "beast" fit well within this description of Satan? Someone who can't die, because he also has the implant, which he forces everyone to get, or they are beheaded. A non-human entity that people will worship. A being that declares himself to be God and demands we all worship him. if the Rapture happens after they are disclosed, Christians will also be deceived by him. If we are all gone, he will have a full reign of terror on the Earth, and nothing will stop him until the final battle of Armageddon!

Many, at this point in this book are still not convinced these creatures are demons, demonic, or even evil. Why can't they be aliens from another faraway planet here to "save us" from ourselves? First of all, Jesus is the only one who can "save" us. These creatures can only condemn us. Look at their modus operandi which is just the same as demons at the beginning; creating creatures that are not human, but corrupted. It was giants before, now just more human. The end result will be the same, corrupt creation by making demonic creatures. And then, Satan rules the corrupt Earth. They tried it before, and God stopped it with the Flood. Noah and his family must have been the last few people alive that were not corrupted. This time with the Rapture, most likely Christians will be raptured out before all these things take place, but we still may experience trials before that happens. The Demonic Alien Overlords may show their hand before the Rapture occurs to try to gain the upper hand. For sure, Satan plans to do whatever he can to gain control of us, so be prepared for what happens next.

THE TRAVIS WALTON EXPERIENCE

I mentioned this case before, but it deserves a deeper dive into what happened and how it too correlates with the agenda of these Demonic Alien Overlords and how they plan to take over the world. In his experience, the

Gray alien creatures were there, but also very human-looking creatures were also aboard the craft. The article "Travis Walton Abduction – 1975" in *MUFON Mutual UFO Network* reports that:

… Just then, Walton heard a sound behind him. He turned, expecting more of the short, large eyed creatures, but was pleasantly surprised to see a tall human figure wearing blue coveralls with a glassy helmet. At the time, Walton said, he did not realize how odd the man's eyes were: larger than normal, and a bright gold color.

Walton says he then asked the man a number of questions, but the man only grinned and motioned for Walton to follow him. Walton also said that because of the man's helmet he might have been unable to hear him, so he followed the man down a hallway which led to a door and a steep ramp down to a large room Walton described as similar to an aircraft hangar. Walton says he realized he had just left a disc-shaped craft similar to the one he had seen in the forest just before he had been struck by the bluish light, but the craft was perhaps twice as large.

In the hangar-like room, Walton reported seeing other disc-shaped craft. The man led him to another room, containing three more humans — a woman and two men — resembling the helmeted man. These people did not wear helmets, so Walton says he began asking questions of them. They responded with the same dull grin, and led him by his arm to a small table.[125]

So, as I keep reiterating, the agenda is to produce human-alien-hybrids, which is similar to what was done

---

[125] From *MUFON Mutual UFO Network*, "Travis Walton Abduction – 1975" https://www.mufon.com/travis-walton-abduction---1975.html Accessed 7.17.2021

according to the Scripture back in Genesis chapter 6. Once they develop these creatures they will then be in a position to take over the world, with the corrupted creatures as their "military" force. Think about it, if aliens were living on another planet, would they look almost exactly like us? If evolution is true, would they evolve to be just like us? Even on *Star Trek* or *Star Wars*, they don't. Aliens, if they exist, are going to be totally different than we are. But, if they can beguile the United States Government to assist them with their plans of making alien-human hybrids, these creatures would be far superior to us and would complete the agenda to **rule** us. Right now, the Deep State thinks we are assisting them so they can survive when in reality we are assisting them to destroy us!

You are right, I have absolutely no proof of what I am saying here. But if these creatures are demons, and not actually aliens, I don't need any, because we already have the authority of the Bible to prove what I am saying. Demons aren't nice creatures that help you succeed, but evil, vile oppressors that are intent on destroying everything God made and ruling it themselves. Make no mistake, they will deceive you and everyone else into thinking they are aliens, but everything points to demons.

HAIR OF THE ALIEN – THE PETER KHOURY CASE

I hesitated to put this account in this book, as it is somewhat of a prurient nature, but it needs to be examined because in this case, there is physical evidence of the incident. On Thursday, July 23, 1992, Peter Khoury was awakened by two female aliens that assaulted him in his bed.[126] In this instance, he was assaulted by two non-human alien females, a blond and a Chinese girl and he records how he found a hair under his foreskin.[127] From an account in his diary:

Thursday July 23, 1992:

Had a very weird experience this morning with two females. 7.00 a.m. after dropping Viv off at station got really sick a few times went straight to bed. All of a sudden two females appeared from nowhere. A blond and a Chinese girl. Very weird looking eyes and color of blond. Both were naked. Blond pushed me to her breast then I bit her nipple and started coughing. She showed no emotion, blood or screamed in pain. I went to toilet and found two hairs under my foreskin. I put them in plastic bag. It is in my filing cabinet.[128]

This account would be nothing but a weird dream if it were not for the analysis of the hair he had in his possession. When analyzed the hair is not human, and not man-made or of some other known creature, but *alien!* The subsequent scientific DNA analysis yields these strange results:

"The blonde alien hair revealed an extraordinary anomaly. Depending on whether we analyzed the hard hair shaft or the soft root, its mitochondrial DNA

---

[126] From <u>Hair of the Alien DNA and Other Evidence for Alien Abductions,</u> by Bill Chalker, Paraview Pocket Books 2005 p. 24 Google Books
https://books.google.co.uk/books?id=hMh5hBuKZhMC&pg=PP6&source=gbs_selected_pages&cad=2#v=onepage&q&f=false   Accessed 7.18.2021
[127] Ibid p. 25
[128] Ibid p. 25

appeared to be of two different kinds. From the lower hair shaft we again obtained a rare Chinese mitochondrial DNA substitution. But from soft root tissue, we obtained a novel Basque/Gaelic type mitochondrial DNA, which had a rare substitution for that racial grouping along with several other characteristic substitutions.

This in itself was a stunning result. The testing methodology meant that prosaic explanations such as contamination or laboratory error were ruled out. In any normal human DNA, we should obtain consistent DNA irrespective of where the sample comes from, be it hair, blood, or other tissue. The biochemists could not explain this strange anomaly. There was no evidence of a somewhat rare DNA phenomenon called heteroplasmy (where two different mitochondrial DNAs rarely appear within the same sample, usually a result of coexistence of mutant mitochondrial and "wild type" DNA molecules within a cell or tissue). Heteroplasmy, which is more readily found in human hair than other parts of the body, refers to single base transitions in the mitochondrial DNA. For example, G to A, or C to T. They are not big changes. Environmental exposure and aging can be factors. A research article in Nature Biotechnology in 2000 that described cutting edge hybrid cloning techniques to treat hair loss provided a clue. We may have encountered evidence of an extraordinary alien analogue of these techniques in Khoury's encounter. Perhaps even more controversially, we also have findings of nuclear DNA suggestive of possible viral resistance to HIV-AIDS for example – referred to as the CCR5 deletion factor. The implications are startling because less than 1 % of the population has this deleted CCR5 factor, which makes the already unusual hair sample even more provocative. And the CCR5 mutation occurred only about 5,000 years ago, further adding to the intrigue. Still, I must note that the limited nuclear DNA results were insufficient to achieve a completely clear result on this matter."[129]

As we examine what happened here, we can conclude that in this case, two demonic alien entities were following the same modus operandi as before, except this time it was not male fallen angels (demons), mating with females, but two female (human-alien demonic entities) that were mating with a human male to obtain the result of humanoid-alien-hybrids. Although Peter does not remember feeling sexually excited during this episode[130] it is clear the objective here was to commit an act that would result in obtaining genetic material at least. If both were impregnated, then at least two more human-alien-hybrids would be produced. It is not clear if they succeeded in their mission or not, but we can surmise from their actions what they meant to do. Given this story seems outlandish, because there is evidence to support the account, we have to conclude that things like this are actually happening now, and in greater frequency than in the past. True highly evolved aliens would not need to conduct themselves in this manner, I submit, but demons would do exactly what these demons did. Think about what true aliens would do. They have the technology to travel all the way from another solar system, perhaps billions of miles away from Earth, and come down here to

---

[129] "Hair of the Alien- The DNA Paradigm," By Bill Chalker, 2005 http://auforn.com/Bill_Chalker_35.htm Accessed 7.18.2021

[130] From Hair of the Alien DNA and Other Evidence for Alien Abductions, by Bill Chalker, Paraview Pocket Books 2005 p. 36 Google Books
https://books.google.co.uk/books?id=hMh5hBuKZhMC&pg=PP6&source=gbs_selected_pages&cad=2#v=onepage&q&f=false Accessed 7.18.2021

mate with humans! Come on! They wouldn't need or want to mate with us, but demons would! Alien females would have alien males to mate with, and if they mated with us somehow, that would be even worse than if a human were to try to mate with an animal. Such a thing could not be done. For demonic-human hybrids, however, that could be a starting point on the way to the main objective. Total world domination by Demonic Alien Overlords!

Further proof that these beings are not aliens but demons is they are practicing sexual depravity when they either use our genetic material or in fact "mate" with us. The Bible calls this adultery or in a more concise term, fornication. Any sex for a man outside of marriage is against the laws of God. These creatures, even if only aliens, have no reverence for God if they have a God and bring no good to us when they are using us for a sexual purpose, even if they are only gathering genetic material for propagating their continued existence by developing alien-human hybrids. The Bible gives ample commands to not engage in sex outside of the marriage between one man and one woman within the bonds of marriage. Here are the main verses that command us not to do it:

> **Thou shalt not commit adultery.** Genesis 20:14
> **But I say unto you, That whosoever looketh on a woman to lust after her hath committed adultery with her already in his heart.** Matthew 5:28

It is clear from these simple verses that outside of the marriage covenant, between one man and one woman, sex or even the lusting after someone, woman or man, is prohibited.

Now I am not some prude that is "holier than thou!" I know that every man is guilty of this sin, either in the actual act or the act of lusting after women, and I bet all women are also guilty of lusting after men. I am guilty, as is every person who is of age. So, if we look at what is going on with these "aliens" we can deduce from their actions, they are demonic. Again, the Bible gives us the way to know what is truly good and what is absolute evil:

> **Ye shall know them by their fruits. Do men gather grapes of thorns, or figs of thistles?** Matthew 7:16

We see by their actions are that they are demonic entities, not highly evolved aliens. But we still have scientists hoping for all their lives there is some alien race out there that will come and "save" us by providing highly advanced technology and free energy to make our world a better place. If we watch any alien horror movies, we know that is not the case. In the movie *Species*[131] we were being offered advanced technology and how to blend our DNA with theirs to produce a "super-human", but it backfired and if they would have succeeded, our world would have been theirs! One Scripture still reigns true when we look at what is happening with these UFOs:

> **And ye shall know the truth, and the truth shall make you free.** John 8:32

We must realize the truth before it is too late. Once these creatures arrive on the scene for real, we will have a hard time believing they are demonic. We have to be prepared to expose them and be willing to fight them. When people begin to be healed of their diseases by accepting the implant, at first it will seem miraculous, but then when we find they are becoming the slaves of these aliens, we will be unable to fight.

---

[131] *Species*, 1995, directed by Roger Donaldson.

Look at what happened to this man. Two naked female alien-human hybrid creatures show up at his bed and try to seduce him into having sex with him. Proof of what the goal was is that he found an alien hair under his fore-skin. This is sexual depravity that is demonic.

I don't want to belabor the point, but the world finds this type of salacious behavior enticing. When the aliens are revealed for real, they may even send out hybrid women to seduce men and hybrid men to seduce women. Who knows what the devil might try to do to keep people from believing in God? Get a man interested in sex and you can forget all else. Women would also fall prey to the evil actions of these Demonic Alien Overlords. They are preparing the way now for such a heinous act. Our only hope is that we can resist. On our own strength, we can't. Jesus is again the only answer to resisting this type of debauchery!

Chapter 16
WHO IS THE DEVIL?

Satan is not some mythological creature dreamed up in our minds to explain away evil. He is a real entity, a fallen angel from Heaven that rebelled against God and wants to be God. I only wish it weren't true that Satan exists. Then we could all relax, and this book would be a moot point. If there is no devil, then there could be aliens out there that will arrive to help us. Since Satan exists, then any so-called aliens would be controlled by him, as he is the "god" of this world. We first hear about him in the Bible in Genesis chapter three. He tempts the woman disguised as a serpent and convinces her to eat the forbidden fruit. He lies to the woman and gets her to also doubt God. The account is as follows:

¹ Now the serpent was more subtil than any beast of the field which the LORD God had made. And he said unto the woman, Yea, hath God said, Ye shall not eat of every tree of the garden?

² And the woman said unto the serpent, We may eat of the fruit of the trees of the garden:

³ But of the fruit of the tree which is in the midst of the garden, God hath said, Ye shall not eat of it, neither shall ye touch it, lest ye die.

⁴ And the serpent said unto the woman, Ye shall not surely die:

⁵ For God doth know that in the day ye eat thereof, then your eyes shall be opened, and ye shall be as gods, knowing good and evil.

⁶ And when the woman saw that the tree was good for food, and that it was pleasant to the eyes, and a tree to be desired to make one wise, she took of the fruit thereof, and did eat, and gave also unto her husband with her; and he did eat.

Genesis 3: 1 - 6

Notice the similarities between what was done at the beginning with the devil and man and what is happening now with the UFO phenomena. The things that God said about him creating us aren't really true. We are your "god", or so say the aliens. We can "help" you to overcome death. We have great knowledge that will "save" you. God lied and you won't really die when you take our implant. You will be as gods, knowing good and evil. You will have knowledge that will help you live on the Earth that God is keeping from you. The idea that they will give us their knowledge in exchange for genetic material, is the same old lie with a little twist. The first objective of Satan, though, is to get rid of God! There is no God because we are aliens, and we have evolved and were not created, therefore, we have seeded the Earth, and you were produced! While this is a huge lie, people will be inclined to believe it because the deception will be so advanced.

Satan was, according to Scripture, the chief angel in charge of praising God. It describes his demise in the book of Isaiah.

> [12] How art thou fallen from Heaven, O Lucifer, son of the morning! how art thou cut down to the ground, which didst weaken the nations! [13] For thou hast said in thine heart, I will ascend into Heaven, I will exalt my throne above the stars of God: I will sit also upon the mount of the congregation, in the sides of the north: [14] I will ascend above the heights of the clouds; I will be like the most High. [15] Yet thou shalt be brought down to Hell, to the sides of the pit. [16] They that see thee shall narrowly look upon thee, and consider thee, saying, Is this the man that made the earth to tremble, that did shake kingdoms; [17] That made the world as a wilderness, and destroyed the cities thereof; that opened not the house of his prisoners? Isaiah 14: 12 - 17

He fell because he wants to be God. These demonic creatures, masquerading as aliens will try to pass themselves off as God, with unbelievably advanced technologies beyond our imagination, and the world will blindly accept this explanation. Satan, their leader, a demonic-human-alien-hybrid will pass himself off as the Jewish Messiah! That is what appears to be happening. I could be totally wrong, but just remember, when they are disclosed, they are not aliens, but demons. Look at the Scripture again, he is a "man", but not a normal man.

> **[11]** **And I beheld another beast coming up out of the earth; and he had two horns like a lamb, and he spake as a dragon.**
> **[12]** **And he exerciseth all the power of the first beast before him, and causeth the earth and them which dwell therein to worship the first beast, whose deadly wound was healed.**
> **[13]** **And he doeth great wonders, so that he maketh fire come down from Heaven on the earth in the sight of men,**
> **[14]** **And deceiveth them that dwell on the earth by the means of those miracles which he had power to do in the sight of the beast; saying to them that dwell on the earth, that they should make an image to the beast, which had the wound by a sword, and did live.**
> **[15]** **And he had power to give life unto the image of the beast, that the image of the beast should both speak, and cause that as many as would not worship the image of the beast should be killed.**
> **[16]** **And he causeth all, both small and great, rich and poor, free and bond, to receive a mark in their right hand, or in their foreheads:**
> **[17]** **And that no man might buy or sell, save he that had the mark, or the name of the beast, or the number of his name.**
> **[18]** **Here is wisdom. Let him that hath understanding count the number of the beast: for it is the number of a man; and his number is Six hundred threescore and six.** Revelation 13: 11 - 18

Notice that it describes:

"...another beast coming up out of the Earth." True because their bases are here on the Earth from which they dispatch their evil.

We are going to be deceived because of their power because he can "make fire come down from Heaven." These demonic alien crafts would be able to do just that.

He is going to show up, have his head nearly cut off, and still live because he has the implant. Because he does these wonders he will demand all of the world to get the implant so they too can live "forever" but it will be in Hell. The implant will actually make people slaves to them.

Those that refuse to get the implant will be unable to buy food, meaning they won't be able to eat without it.

He is a "man" but also a "beast." A demonic-human-hybrid fits this bill to a tee!

If I keep rewriting the same words in this book it is only because we tend to gloss over what we hear until it is repeated several times. We don't always hear something once and accept it. Even when people lie, they keep repeating the same lie over and over and then people believe it. I am, therefore, repeating this truth, until it is universally accepted and before it is too late. The secular community will not accept or believe what I am saying here. Deep down all true Christians will know it is true. Non-Christians are always looking for some other way, besides Jesus, to account for our existence. They don't want to be lashed down with the Ten Commandments that limit how they can live their lives, especially when it comes to the subject of sex. They want to be able to live like the devil, even though they will claim he doesn't even exist. These Demonic Alien Overlords, once they are disclosed completely, will give them the freedom they want, while secretly making them the slaves of Satan, and ultimately dragging them into Hell. Let me write it again: **THESE ALIENS ARE NOT ALIENS BUT DEMONS!** If I am wrong, in this country at least, I should have the right to say it, even if it is wrong. You always have the choice of whether to accept it or not.

The news is already being censored without reason. For months they stopped anyone from being able to say that the Covid-19 virus originated in a lab there in Wuhan. Then they found out it most likely did. But whoever stopped people from saying it broke out from the lab, never had to reprint a retraction, or tell anyone they were wrong. Now if you want to say anything negative about the vaccine, they want that censored. Whatever happened to Freedom of the Press? Likewise, they won't let people say the 2020 election was stolen, even though it is coming to light now that actual fraud was discovered in Georgia. There are also anomalies in Arizona. What if they discover that Trump really won? I say this now because when these "aliens" land at the White House, they won't let you say they are demons, and if you do, you will be arrested on site!

But who is Satan? We don't know yet. I believe he is alive today in the form of a man, (human-alien-hybrid) at one of the bases there in either Area 51 or Dulce, New Mexico, or somewhere else. He is waiting for the right time to reveal himself. Like a farmer waiting to pick the fruit, he is waiting for the world to be "ripe for the picking", and we don't have much time left. The government is releasing information about these so-called aliens in stages, and once the world can accept it, then they will be fully disclosed. With all the lies, the devil tells, he only tells a little of it at one time, until he finally says the whole lie. We are a frog boiling in a pot, and we don't jump out because we haven't got hot enough yet. When we finally realize our plight, it will be too late. Like an army getting into a position to strike, they are getting into a position for taking over the world.

Many people doubt the existence of Satan. If there is no devil, then there is also no God. I am going to "cheat" just a little and rewrite something I wrote in a previous book because it shows how Satan does exist, and therefore God. The last chapter of <u>Finding Proof of Jesus</u> makes it clear that Satan does exist.

Chapter 17

THE BACKWARDS PROOF OF THE EXISTENCE OF GOD

The biggest problem with the theory of evolution is that it denies the spiritual. If man can be said to have evolved from lesser animals that do not have spiritual tendencies, then man must also not have any of these tendencies. But what is true is that man is the only creature that prays. Evolution fails to explain why this would be so.

The Bible, however, makes it very clear as to why this would be so—in that man is the only creature that was created in the image and likeness of God. As it is written in Genesis 1: 26 – 27:

"And God said, Let us make man in our image, after our likeness: and let them have dominion over the fish of the sea, and over the fowl of the air, and over the cattle, and over all the Earth, and over every creeping thing that creepeth upon the Earth.

So God created man in his own image, in the image of God created he him; male and female created he them."[132]

The theory of evolution leaves this explanation out and completely omits any mention of God as Creator.

But even if evolution were true, it cannot remove or ignore God! For if any man says, "There is no God, or I don't believe in God, or I am not sure if there is a God or not," then that man is truly ignorant and has disproven his own statement. Of course in order to say: "There is no God," you have to absolutely be God! If you say there is no God, you have to be in possession of at least two of

---

[132] Holy Bible, King James Version, in public domain, Genesis 1: 26 – 27.

the attributes of God, **Omnipresence and Omniscience.** That is, you have been everywhere all of the time, including now, and you know everything. If there were any place in the entire universe that you do not know of, and are not at now, God could be there; if there were anything that you do not know of, that could be God! You make yourself out to be God—the very thing you said did not exist!

But let us examine the question of God and his existence in another light, which I call: "The Backwards Proof of the Existence of God." What does exist is the Bible, which has been said to be the VERY WORD OF GOD. Contained within it are the Ten Commandments, which are moral absolutes or absolute truths. Now, you may say, "I don't follow them, or I don't believe in them." But even if you do not follow them or believe in them, <u>they still exist</u>!

<u>THEY STILL EXIST!</u>
<u>THEY STILL EXIST!</u>
Even if you choose to not believe in God, which does not change the fact of His existence. If the Ten Commandments are not the absolute truths for you, you still have moral obligations or moral absolutes, as your conscience dictates to you what is right and what is wrong. Just by you saying, "I do not believe in these Ten Commandments," you in effect make yourself out to be God— "The Determiner of Truth."

TEN COMMANDMENTS
I
I AM THE LORD THY GOD ...
THOU SHALT HAVE NO OTHER GODS BEFORE ME*
II

THOU SHALT NOT MAKE UNTO THEE ANY GRAVEN IMAGES

III

THOU SHALT NOT TAKE THE NAME OF THE LORD THY GOD IN VAIN

IV

REMEMBER THE SABBATH DAY TO KEEP IT HOLY

V

HONOR YOUR FATHER AND YOUR MOTHER

VI

THOU SHALT NOT KILL

VII

THOU SHALT NOT COMMIT ADULTERY

VIII

THOU SHALT NOT STEAL

IX

THOU SHALT NOT BEAR FALSE WITNESS AGAINST THY NEIGHBOR

X

THOU SHALT NOT COVET THY NEIGHBORS GOODS

EXODUS 20:1-17

ACCORDING TO ANCIENT SCRIPTURE, THE HEBREWS WERE CAST OUT OF EGYPT BY PHARAOH AND MADE TO WANDER THROUGH THE DESERT IN SEARCH OF THE PROMISED LAND. GOD SPOKE TO THE PROPHET MOSES ON MOUNT SINAI IN THE EGYPTIAN DESERT, WHERE HE ORDERED MOSES TO TAKE HIS COMMANDMENTS TO HIS PEOPLE, SO THEY COULD LIVE ACCORDING TO HIS WISHES.

THESE LAWS FORM THE BASIS OF THE MODERN RELIGIONS OF JUDAISM AND CHRISTIANITY. [133]

*In the First Commandment, it is no other gods besides the God of the Jews, which is **<u>Three in One; God the Father, God the Son, and God the Holy Spirit</u>**, which was proclaimed by the person of Jesus Christ.

Let's say you refuse to accept these as the Absolute Truths of God for your life and feel that you are able, or qualified, to make up your own rules—whatever you want; all of these, or none of these, however, it suits you, because You are God! Not only that, but whenever you sin, you have changed the truth of these laws to suit your situation. In doing so, you have made yourself out to be a determiner of truth. Evolution allows you to make up your own rules; it is survival of the fittest, kill or be killed, steal or be stolen from, commit adultery anytime you can lie to anyone to obtain what you want, etc. Nothing is wrong or right.

If we allow evolution to be the determiner of our truth, we have absolute chaos and a world without laws. You could not complain if someone steals your car, empties your 401K, runs off with your wife, or even kills you and laughs while they do it because that is survival of the fittest! In this type of existence, it is as the Bible makes clear in Proverbs 12:15: *"The way of a fool is right in his own eyes: but he that hearkeneth unto counsel is wise."*[134] I will make the contention here that you have moral and ethical obligations, and these moral and ethical obligations are the God that you say does not exist. Further, I would ask you a question, "Does love exist?" Few people would be so crass as to say love does not exist; not

---

[133] Ten Commandments in public domain.
[134] Holy Bible, King James Version, in public domain, Proverbs 12:15.

the love of a mother for a child, the love of a husband for a wife, the love of a child for their parents; love of a boy for a girl, and so on. Since the case is certainly made for the existence of love, that is God. The Bible says that God is love in innumerable places, including John 4:8:

"He that loveth not knoweth not God; for God is love."[135]

So, if you even have a smidgen of an inkling that love exists, you know deep down inside that God exists as well.

A refusal to acknowledge, much less to obey, the Ten Commandments, is a refusal to acknowledge God—to usurp God, if you will, from His rightful place of worship and praise and enthrone yourself in that position. Man wants to be God. Man in his sinful state does not want Jesus Christ and His rules to have authority in his life. This is the same basic sin that was committed at the very beginning when Satan fell from grace, which is told of in Isaiah 14:12-14:

"How art thou fallen from Heaven, O Lucifer, son of the morning! How art thou cut down to the ground, which didst weaken the nations!

For thou hast said in thine heart, I will ascend into Heaven, I will exalt my throne above the stars of God: I will sit also upon the mount of the congregation, in the sides of the north:

I will ascend above the heights of the clouds; I will be like the most High."[136]

This sinful nature, this refusal to abide by the rules, and the desire to set yourself up as The Determiner of Truth, is proof that all of mankind—including you and I

---

[135] Holy Bible, King James Version, in public domain, I John 4:8.
[136] Holy Bible, King James Version, in public domain, Isaiah 14:12-14

—are "possessed" by the very same spirit as Satan. Here, then, is the primary hypothesis of what I call "<u>The Backwards Proof for the Existence of God." Since Satan can be shown to exist by the sinful nature of all of mankind, God can be shown to exist. Without God existing, neither could Satan. By definition, God is the Creator of all things, and that includes Satan.</u>

There are many other great philosophical proofs of the existence of God, most notably Rene Descartes' and St. Thomas Aquinas' proofs. Descartes' "Cogito ergo sum," that is, "I think, therefore, I am," ranks among the greatest proofs in that he knew one thing: he had a thought. Something is responsible for that thought, and that something is God!

Aquinas' proof, found in <u>Quinque viae,</u> [137] was "The Argument of the Unmoved Mover," which ranks in my opinion as the single greatest proof of all time for the existence of God. It is simply (although oversimplified here) that motion exists, and that something is responsible for that motion. That something is the Unmoved Mover, and this Unmoved Mover is God! This is only one of five proofs that he used to prove the existence of God.

The Quinque viæ (Latin, usually translated as "Five Ways" or "Five Proofs") are five logical arguments regarding the existence of God summarized by the 13th-century Catholic philosopher and theologian St. Thomas Aquinas in his book Summa Theologica. They are:

the unmoved mover;

the first cause;

the argument from contingency;

---

[137] "Quinque viae" Wikipedia, the free encyclopedia, (last updated 8.12.2015), Accessed 09.06.2015 en.wikipedia.org/wiki/Quinque_viae.

the argument from degree;

the teleological argument ("argument from design").[138]

But if the reader has made it to this point in this book and is still unconvinced of the necessity of God and His existence, there is still much more evidence regarding the Bible and the facts surrounding it. I have discussed repeatedly how the theory of evolution fails miserably when mathematics is applied to it. When mathematics is applied to biblical Scripture, however, the situation is much different. Some 2,000 or more biblical prophecies have been literally fulfilled. There are approximately some 333 prophecies that have been literally fulfilled by Jesus Christ. Josh McDowell reports in "Evidence That Demands a Verdict" that there are several hundred references to the coming Messiah.[139] There are 61 main biblical prophecies concerning Jesus Christ. They are listed in the chart below.

---

[138] "Quinque viae" Wikipedia, the free encyclopedia, (last updated 8.12.2015), Accessed 09.06.2015 en.wikipedia.org/wiki/Quinque_viae.

[139] Josh McDowell, Evidence that Demands a Verdict, San Bernardino, Here's Life Publishers, 1979, p. 141.

LISTED EVENT/ACTION	PROPHECY SCRIPTURE	FULFILMENT SCRIPTURE
1. BORN OF SEED OF WOMAN	GEN. 3:15	GAL. 4:4/MAT. 1:20
2. BORN OF A VIRGIN	IS. 7:14	MAT. 1:18, 24,25/LUKE 1:26-35
3. SON OF GOD	PS. 2:7/I CH. 17:11-14/II SAM. 7:12-16	MAT. 3:17/MAT. 16:16/MK. 9:7/LUKE 9:35;22:70/ACTS 13:30-33/JOHN 1:34-49
4. SEED OF ABRAHAM	GEN. 22:18;12:2,3	MAT. 1:1/GAL. 3:16
5. SON OF ISAAC	GEN. 21:12	LUKE 3:23, 34/MAT. 1:2
6. SON OF JACOB	NUM. 24:17	LUKE 3:23, 34/MAT. 1:2/LUKE 1:33
7. TRIBE OF JUDAH	GEN. 49:10/MIC. 5:2	LUKE 3:23/MAT. 1:22/HEB. 7:14
8. FAMILY LINE OF JESSE	ISAIAH 11:1	LUKE 3:23, 32/MAT. 1:6
9. HOUSE OF DAVID	JER. 23:5/II SAM. 7:12-16 PS. 132:11	LUKE3:23, 31/MAT. 1:1; 9:27; 15:22; 20:30, 31;

		21:9, 15; 22:41-46 MARK 9:10; 10:47,48/LUKE 18:38, 39/ACTS 13:22, 23/REV. 22:16
10. BORN AT BETHLEHEM	MICAH 5:2	MAT. 2:1/ JOHN 7:42/MAT. 2:4-8/LUKE 2:4-7
11. PRESENTED WITH GIFTS	PSALMS 72:10/ISAIAH 60:6	MAT. 2:1, 11
12. HEROD KILLS CHILDREN	JER. 31:15	Mr. 2:16
13. HIS PRE-EXISTENCE	MICAH 5:2/ISAIAH 9:6, 7; 41:4; 44:6; 48:12 PSALMS 102:25/PRO. 8:22, 23	COL. 1:17/JOHN 1:1, 2;8:58; 17:5, 24/REV. 1:17; 2:8; 22:13
14. HE SHALL BE CALLED LORD	PSALMS 110:1/JER. 23:6	LUKE 2:11/MAT. 22:43-45
15. SHALL BE IMMANUEL	ISAIAH 7:14	MAT. 1:23/LUKE 7:16
16. SHALL BE A PROPHET	DEUT. 18:18	MAT. 21:11/LUKE 7:16/ JOHN 4:19;6:14;7:40
17. PRIEST	PSALMS 110:4	HEB. 3:1; 5:5, 6
18. JUDGE	ISAIAH 33:32	JOHN 5:30 II TIM. 4:1

19. KING	PSALMS 2:6/ZAC. 9:9 JER. 23:5	MAT. 27:37, 21:5/JOHN 18:33-38
20. SPECIAL ANOINTMENT OF HOLY SPIRIT	ISAIAH 11:2; 42:1; 61:1, 2/PSALMS 45:7	MAT. 3:16, 17; 12:17-21/MARK 1:10, 11/LUKE 4:15-21, 43/JOHN 1:32
21. HIS ZEAL FOR GOD	PSALMS 69:9	JOHN 2:15-17
22. PRECEDED BY MESSENGER	ISAIAH 40:30/MAL. 3:1	MAT. 3:1, 2; 3:3; 11:10/JOHN 1:23/LUKE 1:17
23. MINISTRY TO BEGIN IN GALILEE	ISAIAH 9:1	MAT. 4:12, 13, 17
24. MINISTRY OF MIRACLES	ISAIAH 35:5, 6a; 32:3, 4	MAT. 9:35; 9:32, 33; 11:4-6/MARK 7:33-35/JOHN 5:5-9; 9:6-11; 11:43, 44, 47
25. TEACHER OF PARABLES	PSALMS 78:2	MAT. 13:34
26. HE WAS TO ENTER THE TEMPLE	MAL. 3:1	MAT. 13:34
27. HE WAS TO ENTER JERUSALEM ON A DONKEY	ZACH. 9:9	LUKE 19:35, 36, 37a/MAT. 21:6-11

28. "STONE OF STUMBLING" TO THE JEWS	PSALMS 118:22/ISAIAH 8:14; 28:16	I PET. 2:7/ ROMANS 9:32, 33
29. "LIGHT" TO THE GENTILES	ISAIAH 60:3; 49:6	ACTS 13:47, 48a; 26:23; 28:28
30. RESURRECTION	PSALMS 16:10; 30:3: 41:10; 118:17/ HOSEA 6:2	ACTS 2:31; 13:33/LUKE 24:46/MARK 16:6/MAT. 28:6
31. ASCENSION	PSALMS 68:18a	ACTS 1:9
32. SEATED AT THE RIGHT HAND OF GOD	PSALMS 110:1	HEB. 1:3/ MARK 16:19/ACTS 2:34, 35
33. BETRAYED BY A FRIEND	PSALMS 41:9; 55:12-14	MAT. 10:4; 26:49, 50/ JOHN 13:21
34. SOLD FOR 30 PIECES OF SILVER	ZACH. 11:12	MAT. 26:15; 27:3
35. MONEY TO BE THROWN IN GOD'S HOUSE	ZACH. 11:13b	MAT. 27:5a
36. PRICE GIVEN FOR POTTER'S FIELD	ZACH. 11:13b	MAT. 27:7
37. FORSAKEN BY HIS DISCIPLES	ZACH. 13:7	MARK 14:50; 14:27/MAT. 26:31
38. ACCUSED BY FALSE WITNESSES	PSALMS 35:11	MAT. 26: 59, 60

39. DUMB BEFORE ACCUSERS	ISAIAH 53:7	MAT. 27:12
40. WOUNDED AND BRUISED	ISAIAH 53:5/ZACH, 13:6	MAT. 27:26
41. SMITTEN AND SPIT UPON	ISAIAH 50:6/MICAH 5:1	MAT. 26:67/LUKE 22:63
42. MOCKED	PSALMS 22:7, 8	MAT. 27:31
43. FELL UNDER THE CROSS	PSALMS 109:24, 25	LUKE 23:26/MAT. 27:31, 32
44. HANDS AND FEET PIERCED	PSALMS 22:16/ZACH. 12:10	LUKE 23:23/JOHN 20:25
45. CRUCIFIED WITH THIEVES	ISAIAH 53:12	MAT. 27:38/MARK 15:27, 28
46. MADE INTERCESSION FOR HIS PERSECUTORS	ISAIAH 5:12	LUKE 23:34
47. REJECTED BY HIS OWN PEOPLE	ISAIAH 53:3/PSALMS 69:8; 118:22	JOHN 7:5, 48; 1:11/MAT. 21:42, 43
48. HATED WITHOUT A CAUSE	PSALMS 69:4/ISAIAH 49:7	JOHN 15:25
49. FRIENDS STOOD AFAR OFF	PSALMS 38:11	LUKE 23:49/MARK 15:40/MAT. 27:55, 56

50. PEOPLE SHOOK THEIR HEADS	PSALMS 109:25/PSALMS 22:7	MAT. 27:39
51. STARED UPON	PSALM 22:17	LUKE 23:35
52. GARMENTS PARTED AND LOTS CAST	PSALMS 22:18	JOHN 19: 23; 24
53. TO SUFFER THIRST	PSALMS 69:21; 22:15	JOHN 19:28
54. GALL AND VINEGAR OFFERED HIM	PSALMS 69:21	MAT. 27:34/JOHN 19:28, 29
55. HIS FORSAKEN CRY	PSALMS 22: 1a	MAT. 27:46
56. COMMITTED HIMSELF TO GOD	PSALMS 31:5	LUKE 23:46
57. BONES NOT BROKEN	PSALMS 34:20	JOHN 19:33
58. HEART BROKEN	PSALMS 22:14	JOHN 19:34
59. HIS SIDE PIERCED	ZACH. 12:10	JOHN 19:34
60. DARKNESS OVER THE LAND	AMOS 8:9	MAT. 27:45
61. BURIED IN RICH MAN'S TOMB	ISAIAH 53:9	Mat. 27:57-60

This list of scripture references is excerpted from "Evidence That Demands a Verdict" by Josh McDowell, © 1972 and 1979, Campus Crusade for Christ, published by Here's Life Publishers pages 135- 138. Used with permission.

Critics will contend that these prophecies might have been fulfilled by other men. Yet it could not be possible for one to say the same about all of the 61 biblical prophecies listed here, or the other 332 total. If Jesus Christ fulfilled only eight prophecies, rather than 332 or the 61 main ones here, the chance of any one man being able to fulfill eight would have been 1 in 10^{17} (one in 1,000,000,000,000,000,000) or one in one thousand quadrillion.[140]*

The reliability of the Bible is well-documented and confirmed by historical texts and archaeological evidence. Whenever the Bible can be tested, it is always shown to be accurate; not one bona fide wrong statement can be identified in the Bible from archaeological evidence. For many years, for example, the existence of the cities of the plain near where Abraham lived were thought to be mythological, because they had not been found in any historical references other than the biblical text. Yet, in 1974, the Ebla Tablets were discovered. They mentioned at least four of the cities (Sodom, Gomorrah, and Zoar/Bela), according to Giovanni Pettinato. They traded with the Ebla Kingdom, although some

---

[140] Josh McDowell, Evidence that Demands a Verdict, San Bernardino, Here's Life Publishers, 1979, p. 167, From Peter Stoner in <u>Science Speaks</u>.

*This number is stated to be one in one thousand quadrillion, but it is so large, I don't know what it is, and it doesn't make any difference!

archaeologists may now try to dispute these findings.[141] Thus the Bible is once again proven factual and man in his limited knowledge is proven to be untrue, as declared in Romans 3:4a "...*let God be true, but every man a liar* ... [142]

At least one person has tried to challenge the Bible in court, hoping to collect the cash reward offered to anyone who can find one scientific error in it. In November of 1939, a Mr. William Floyd of New York brought a suit against the Science Bureau that offered the reward, headed by the late Dr. Harry Rimmer; however, the Honorable Benjamin Shalleck of the Fourth District Municipal Court ruled in favor of the Bible. To date the reward of $1,000.00 is still unclaimed.[143]

Many people have chosen to disregard the Bible without studying it and to accept the theory of evolution just because someone else has said that evolution is correct, and the Bible is mythological. Let me suggest that you should make up your own mind about the Bible and the theory of evolution by studying the Scriptures to determine what is truth, rather than accepting blindly the statements of a few uninformed sources. What you have to gain is knowledge about the Bible and the salvation of your soul.

CONCLUSION FOR CHAPTER 17

If the theory of evolution is correct and the Bible is wrong, I have lost nothing by believing in the Bible. I can

---

[141] "Ebla-Biblical Controversy," Wikipedia, the free encyclopedia, (last updated 8.12.2015), Accessed 09.06.2015, en.wikipedia.org/wiki/Ebla-Biblical_controversy.

[142] Holy Bible King James Version, in public domain, Romans 3:4a.

[143] Edward F. Blick, Correlation of the Bible and Science, Southwest Radio Church edition, Oklahoma City, OK, 1976, p. 3.

gain a high quality of life by trusting in the Bible. I have hope beyond the grave. I will live a long and full life, as I am at the age of 67. I have found the answers to the questions of life by believing in God. He has even spoken to me in an audible voice, I believe. I also saw an image of Jesus Christ in a mirror that delivered me from demonic oppression. I am His child and I have an inheritance of blessings that would fill a book with thousands of pages. I am very blessed! If in the end, there is no God, no Heaven, and no Hell—if when we die, we just go six feet under and that is it—I would still choose to believe, for the reasons I mentioned above.

But now, let's turn it around and suppose for a moment that the Bible is true, and there is a God and evolution is wrong. What have you lost if you do not believe? Everything, including the salvation of your soul. Then you are destined to go to an eternal Hell with torments forever, while I will go to an eternal Heaven with blessings forever.

Please pray this prayer out loud if you want to believe in Jesus and not go to an eternal Hell:

"Dear Jesus, I repent of my sin and ask you to come into my heart and save me from my sin. I believe in you and believe that you died on the cross and rose from the dead to save me from my sin. I make you Lord and Savior of my life and I invite you to come into my heart and save me! Help me to live for you. In Jesus' name, Amen."

If you have prayed this prayer, the Bible says you are saved, as it says in Romans 10:9, 10:

"That if thou shalt confess with thy mouth the Lord Jesus, and shalt believe in thine heart that God hath raised him from the dead, thou shalt be saved.

For with the heart man believeth unto righteousness; and with the mouth confession is made unto salvation."[144]

Once you have accepted Jesus, begin going to a Bible-believing church and read the Bible every day. Tell someone you have accepted Jesus. Be baptized as soon as possible.

This concludes the bulk of what I wrote in the final chapter of my book <u>Finding Proof of Jesus,</u> and I am re-printing it here because if Satan can be shown to exist, then God has to exist, and that is The Backwards Proof for the Existence of God. Basically, when these demons and Satan arrive on the scene, disguised as aliens, just remember **THEY ARE DEMONS AND NOT ALIENS.**

The conclusion that can be reached in this chapter is that ample evidence exists for Satan and demons to exist, but as yet there is no conclusive proof that aliens exist. Satan and his demons would "masquerade" as aliens, thus trying to get the world to follow them.

---

[144] Holy Bible King James Version, in public domain, Romans 10: 9, 10."

Chapter 18
PROOF OF THE GLOBAL CATASTROPHIC FLOOD AND THEREFORE DISPROOF OF EVOLUTION AND ALIENS

Most people you ask today will say they don't believe in the fact of the Global Flood in Genesis. The evidence for a World-wide Global Catastrophic Flood is over-whelming when we look at the geological records. I bring this up because since there was a global flood, then the Earth is young, and aliens would not have had time to evolve, period! This would mean that unless God created them, they don't exist. Since he didn't tell us about them in the Bible, he didn't create them, therefore, they don't exist!

God did this act to stop the corruption of the line to Jesus. If there had not been a flood, then there could have not been a Messiah. Noah and his family members were evidently the only people that had not been cor-rupted by the devil and his demons.

Despite the fact you can actually visit the archeologi-cal site where the Ark was found, few will still acknowledge the fact of the Flood. It is where it said it would be in the Bible! It is the same size it said it would be. There is now a national park in Turkey! No denying it, it is a boat the size and at the location, the Bible says it would be! Here is the bulk of what I wrote in my pre-vious book <u>Finding Proof of Jesus</u> that shows insur-mountable evidence for the flood.

A second insurmountable hurdle for the theory of evo-lution is the geological record of the universal flood. In nearly every ancient culture, the flood story is told, and geological evidence supports it. Flood accounts are found in Persia, Syria, Asia Minor, Greece, Egypt, Italy, Lithuania, Wales, Scandinavia, Lapland, Russia, China, India, Central America, Pacific Islands, and even North American cultures.[145] North American cultures may have as many as 37 or more accounts.[146] The excellent book The Deluge Story in Stone, by Byron C. Nelson, Th.M., provides more detail.

The Sumerians and the Babylonians brought us the well-known "Epic of Gilgamesh" and of course the Hebrew people brought us the most well-known account in the Book of Genesis, which, as Josh McDowell in his Evidence That Demands a Verdict has stated, *"The Bible is trustworthy and historically reliable."*[147]

Most evolutionists do not deal with a universal flood construct but suggest that the geological formations have been formed by gradual changes (uniformity).

PLATE TECTONICS

It is theorized that plate tectonics beneath the ocean floor is responsible for the formation of the mountain ranges. It is supposed that these plates "pushed" the mountains upward. Many millions, perhaps billions of years (it is supposed) would be required to form the geological formations we now have. Processes of erosion

---

[145] Byron C. Nelson, Th.M., The Deluge Story in Stone. Minneapolis, Minnesota. Bethany Fellowship, Inc. Publisher, 1968, p. 170-190.

[146] From the film, "The World That Perished," Christian Answers.Net.

[147] Josh McDowell, Evidence That Demands a Verdict Historical Evidence for the Christian Faith. San Bernardino, Here's Life Publishers, p. 74.

would also tend to break them down. Volcanic action is also a factor. One reason why this process of plate tectonics is not entirely responsible for the many mountain formations we now have is, of course, the factor of erosion, which in my opinion would tend to wear down any slow changes as fast as they were formed. More importantly, the uniformitarianism construct is not well supported by major geologic processes such as sedimentation, erosion, deposition, volcanism, glaciations, diastrophism, etc.[148]

A uniformity construct would require huge amounts of time to produce the mountain formations. Yet, John C. Whitcomb and Henry M. Morris state in their book The Genesis Flood/The Biblical Record and its Scientific Implications that many of the mountain ranges have been formed over a relatively short period of time. They quote R. F. Flint from Geological Geology and the Pleistocene Epoch:

"In North America late Pliocene or Pleistocene movements involving elevations of thousands of feet are recorded in Alaska and in the coast ranges of Southern California ... In Europe, the Scandinavian Mountains were created from areas of very moderated relief and conspicuously uplifted in Pleistocene and late Pleistocene time. In Asia there was great early Pleistocene uplift in Turkestan, the Pamir, the Caucasus, and central Asia generally. Most of the vast uplift of the Himalayas is ascribed to the "latest Tertiary" and Pleistocene. In South America the Peruvian Andes rose at least 5,000 feet in post-Pliocene time ... In addition to these tectonic movements many of the high volcanic cones around the Pacific

---

[148] John C. Whitcomb and Henry M. Morris, The Genesis Flood/The Biblical Record and its Scientific Implications. Grand Rapids; Baker Book House, 1961, pp. 124 and following.

border in western and central Asia and in eastern Africa are believed to have been built up to their present great heights during the Pliocene and Pleistocene."[149]

Whitcomb and Morris continue:

"Since the Pliocene and Pleistocene are supposed to represent the most recent geological epochs, except that of the present, and since nearly all of the great mountain areas of the world have been found to have fossils from these times near their summits, there is no conclusion possible other than that the mountains (and therefore the continents of which they form the backbones) have all been uplifted essentially simultaneously and quite recently. Surely this fact accords well with the biblical statements."[150]

If a uniformity construct is considered, then huge amounts of time would be required for the formation of mountain ranges. Since the evidence indicates a relatively short period of time, the biblical account of the flood and a catastrophic construct is supported.

Only the main problems will be discussed here. As Whitcomb and Morris further state:

"But as a matter of fact it is not even true that uniformity is a possible explanation for most of the Earth's geologic formations, as any candid examination of the facts ought to reveal."[151]

---

149 Whitcomb and Morris, quoting R. F. Flint. Glacial Geology in the Pleistocene Epoch. New York, Wiley, 1947, p. 128.

150 John C. Whitcomb and Henry M. Morris, The Genesis Flood/The Biblical Record and its Scientific Implications. Grand Rapids; Baker Book House, 1961, p. 128.

151 John C. Whitcomb and Henry M. Morris, The Genesis Flood/The Biblical Record and its Scientific Implications. Grand Rapids; Baker Book House, 1961, p. 137.

Whitcomb and Morris demonstrate why gradual changes alone cannot be responsible for the mountain ranges with respect to volcanism producing igneous rock formations: *"These igneous rocks are found all over the world in great profusion. Often they are found intruding into previously deposited sedimentary rocks or on the surface covering vast areas of earlier deposits.*

The Columbia Plateau of the northwestern United States covers about 200,000 miles of incredible thickness."[152]

As Whitcomb and Morris further maintain:

"But nothing ever seen by man in the preset era can compare with the phenomena that caused the formation of these tremendous structures. The principle of uniformity breaks down completely at this important point in geologic interpretation. Some manifestation of catastrophic action alone is sufficient."[153]

Whitcomb and Morris further point out sedimentation beds up to thousands of feet thick, fossil graveyards, and the preservation of fossils, raindrops, and even what appear to **be human footprints with dinosaur footprints in the same rock formations.**

The pictures below are photographs of human footprints found in the same rock formations as dinosaur footprints from The Genesis Flood. Human footprints like these are common in these rock formations.

---

[152]John C. Whitcomb and Henry M. Morris, The Genesis Flood/The Biblical Record and its Scientific Implications. Grand Rapids; Baker Book House, 1961, p. 138.

[153] John C. Whitcomb and Henry M. Morris, The Genesis Flood/The Biblical Record and its Scientific Implications. Grand Rapids; Baker Book House, 1961, p. 139.

CONTEMPORANEOUS FOOTPRINTS OF HUMAN AND DINOSAUR

These tracks were both cut from the Paluxy Riverbed near Glen Rose, Texas, in supposedly Cretaceous strata, plainly disproving the evolutionists' contention that dinosaurs were extinct some 70 million years before man "evolved." Geologists have rejected this evidence, however, preferring to believe that the human footprints were carved by some modern artist, while at the same time accepting the dinosaur prints as genuine. If anything, the dinosaur prints look more "artificial" than the human, but the genuineness of neither would be questioned at all were it not for the geologically sacrosanct evolutionary timescale[154]

REPRINTED WITH PERMISSION FROM HENRY M. MORRIS AND JOHN C. WHITCOMB

---

[154] John C. Whitcomb and Henry M. Morris, The Genesis Flood/The Biblical Record and its Scientific Implications. Grand Rapids; Baker Book House, 1961, p. 174

GIANT HUMAN FOOTPRINTS IN CRETACEOUS STRATA.

These are more of the apparently human footprints found in the Paluxy Riverbed. Note the tremendous size, which immediately reminds one of the biblical statement that there were "giants in the Earth in those days" (Genesis 6:4). Similar giant human footprints have been found in Arizona, near Mt. Whitney in California, near the White Sands in New Mexico, and other places.

REPRINTED WITH PERMISSION FROM HENRY M. MORRIS AND JOHN C. WHITCOMB[155]

---

[155] John C. Whitcomb and Henry M. Morris, The Genesis Flood/The Biblical Record and its Scientific Implications. Grand Rapids; Baker Book House, 1961, p. 175

THE FLOOD AS A MOUNTAIN-FORMER

The Biblical account states that *"In the six hundredth year of Noah's life, in the second month, the seventeenth day of the month, the same day were all the fountains of the great deep broken up and the windows of Heaven were opened."*[156] It is evident from this passage that a number of forces can be held responsible for the formation of the mountain ranges:

There was a huge amount of volcanic activity present, which could account for the formation of the huge mountain ranges in conjunction with the flood.

The force of waves deep enough to cover the whole Earth, in conjunction with volcanic activity, would have changed the Earth entirely.

"The force of ocean waves striking the shore can be measured, and has been found to reach three tons per square foot,"[157] remark Whitcomb and Morris, quoting Thompson King. In recent times there has been nothing to compare with the description of the Noachian Deluge as found in the Biblical Scriptures. However, sometimes the destructive power of ocean waves are encountered during storms and Earthquakes, as Whitcomb and Morris report from accounts found in *Scientific American*:

"The great Krakatoa Earthquake, in the East Indies in 1883, created immense waves at least 100 feet high and traveling up to 450 miles per hour, inundating neighboring islands and drowning nearly 40,000 people. A tsunami from the quake was still two feet high as it passed Ceylon and nine inches high at Aden beyond the Arabian Sea! In 1946, a tsunami originating in a quake in the

---

[156] Holy Bible King James Version, in public domain, Genesis 7:11.
[157] Whitcomb and Morris, quoting Thompson King, <u>Water</u>, New York, Macmillan Co., 1953, p.263.

Aleutian Island region traveled 470 miles per hour across the Pacific, creating a 19-foot high "tidal" wave on the shores of Hawaii, with great destruction. A wave that swept across the Bay of Bengal in 1876 left 200,000 people dead."[158]

It is evident from the Scriptures that prior to the Noachian Deluge, mountains and high hills were already present. But what is not determined is how the mountain ranges would have been formed in their present states with many of the fossils embedded within them. A uniformity construct with plate tectonics would not be an adequate explanation, considering issues such as misplaced strata and fossil graveyards.

THE PROBLEM OF MISPLACED STRATA

A huge problem generally not discussed in the textbooks is the fact that nowhere in the world is the "supposed" geological column found. Everywhere, strata is mixed up. Supposed "young" strata are often found beneath supposed "old" strata. An article by M. King Hubbert and Wm. W. Rubey, "Role of Fluid Pressure in Mechanics of Overthrust Faulting," is cited by Whitcomb and Morris:

"But if this is the import of individual creatures which are found out of place in the sequences, what should be said of the many examples of entire formations being out of place in the standard geologic timetable? In every mountainous region on every continent, there seem to be

---

[158] Whitcomb and Morris, quoting Willard Bascom, 'Ocean Waves,' 'Scientific American, vol. 201, August 1959, p.264.

numerous examples of supposedly "old" strata superimposed on top of "young" strata."[159]

While this is not necessarily an example of an overthrust, mountain ranges appear on every continent that will show "old" strata atop "young" strata. [160]

---

[159] Whitcomb and Morris, quoting M King Hubbert and Wm. W. Ruby, "Role of Fluid Mechanics of Overthrust Faulting," 'Bulletin of Geological Society of America' vol. 70 Feb. 1959. P. 180.
[160] Owned image from art explosion

Glacier National Park, Montana. Summit of Chief Mountain, viewed from the northern end of the mountain. Yellow Mountain in the distance. 1911. [161]

Photo credits: Campbell M. R. Collection Glacier National Park andPreserve, National Parks, photo print

US Geological Survey photograph in the public domain.

Chief Mountain photograph from the US Department of Interior in the public domain. This mountain has "old" strata on top of "young" strata.

---

[161] Chief Mountain from the U.S. Geological Survey website in public domain

Lewis Overthrust, which is displayed in The Genesis Flood and also available from the US Geological Survey, which also has "old" strata on top of "young" strata.[162] The U.S.

Geological survey in the public domain.Glacier National Park, Montana. Lewis Overthrust on Summit Mountain, viewed from Marias Pass. 1952. Figure 139, US Geological Survey professional paper 294-K; Figure 21, USGeological Survey Professional paper 296.

Photo credits: Campbell M. R. Collection Glacier National Park and Preserve, National Parks, photo print.

---

[162] Lewis Overthrust from the US Geological Survey website in public domain.

The Colorado River did not form the Grand Canyon, but the Grand Canyon, caused by the great flood, is the cause of the Colorado River.[163]

164

A great example is where we can see the great columns in the Grand Canyon, which shouldshow the geological column, but we will not find the fossils in the order as depicted in thefossil record for the geological records.

---

[163] Owned image from Art Explosion.
[164] Owned image from Art Explosion.

Briefly stated, Whitcomb and Morris examine the tremendous number of formations out of sequence and explain that these could not be possible under a uniformity construct, as Hubert and Ruby admit:

"Since their earliest recognition, the existence of large overthrusts has presented a mechanical paradox that has never been satisfactorily resolved."[165]

Whitcomb and Morris discuss at length many of these examples, including Chief Mountain, Lewis Overthrust, the Matterhorn, etc.:

"To illustrate the character of these important areas, we might consider the well-known Heart Mountain Thrust of Wyoming. This supposed thrust occupies roughly a triangular area, 30 miles wide by 60 miles long with its apex at the northeast corner of Yellowstone Park. It consists of about 50 separate blocks of Paleozoic strata (Ordovician, Devonian, and Mississippian) resting essentially horizontally and comfortably on Eocene beds, some 250,000,000 years younger."[166]

While there is evidence of plate tectonic activity causing numerous geological formations such as "folding, tilting and dike-extruded" areas, etc., the numerous examples of these formations out of sequence cause extreme problems for the uniformity construct in the theory of evolution, and a catastrophic construct is again better supported. Seen in the previous pages are pictures of Chief Mountain and the Lewis Overthrust. Both of these formations are difficult to explain with the uniformity

---

[165] John C. Whitcomb and Henry M. Morris, <u>The Genesis Flood/The Biblical Record and its Scientific Implications.</u> Grand Rapids; Baker Book House, 1961, p. 181.

[166] John C. Whitcomb and Henry M. Morris, <u>The Genesis Flood/The Biblical Record and its Scientific Implications.</u> Grand Rapids; Baker Book House, 1961, p. 187.

construct. For instance, the Lewis Overthrust has a supposed overthrust region some 350 miles wide and six
miles thick, with an inferred horizontal displacement of
at least 35 or 40 miles.

–The Bible describes the Flood in this manner in rela-
tion to the destruction to life on Earth in Genesis 7:21-23:

*"And all flesh died that moved upon the Earth, both of
fowl, and of cattle, and of beast, and of every creeping
thing that creepeth upon the Earth, and every man:*

*All in whose nostrils was the breath of life, of all that
was in the dry land, died.*

*And every living thing was destroyed which was upon
the face of the ground, both man, and cattle, and the
creeping things, and the fowl of Heaven: and they were
destroyed from the Earth: and Noah only remained alive,
and they that were with him in the Ark."*[167]

It is evident from this description that all non-amphib-
ious animal life would have died unless it was on the Ark.
Plants would have remained alive because they germinate both from seeds that germinate well in plain water

---

[167] Holy Bible King James Version, in public domain, Genesis 7:21-
23.

and also from cuttings. According to Whitcomb and Morris, *"The Biblical Deluge is a quite adequate solution."*[168]

The Ark Encounter (Ark built to the dimensions in the biblical text). [169]

NOAH'S ARK

We have all heard the story of Noah's Ark. It is one of the major accounts in Genesis. It is even attested to by Jesus. He states emphatically:

"But as the days of Noah were, so shall also the coming of the Son of man be. [38] For as in the days that were before the flood they were eating and drinking, marrying and giving in marriage, until the day that Noah entered into the ark,

[39] And knew not until the flood came, and took them all away; so shall also the coming of the Son of man be."[170]

Yet, few people will even admit they believe this story. I heard of a professor who asked that everyone close their eyes and raise their hands if they believed the story. Just one student did. He peeked and found that he was the only one who had actually raised his hand.

But the story of Noah's Ark is a more viable explanation for the existence of our current biological construct than evolution could ever be! Keep in mind that in an evolutionary construct, we have to start with **nothing**! No

---

[168]John C. Whitcomb and Henry M. Morris, <u>The Genesis Flood/The Biblical Record and its Scientific Implications.</u> Grand Rapids; Baker Book House, 1961, p. 280.

[169] Owned Image taken 9/2/2021 at The Ark Encounter in Williamstown, KY.

[170] Holy Bible, King James Version, in public domain, Matthew 24: 37-39.

animal life forms, no Earth, no sun, no stars, no matter, no dust particles, no molecules of matter, no atoms of gas, nothing! And then we arrive at everything. Absolutely and unequivocally preposterous!

The Ark has an easily discernible mathematical calculation to determine its viability. The size of the Ark was 437.5 feet long, 72.92 feet wide and 43.75 feet high.[171] This is based on a cubit of 17.5 inches. [172] It also had three decks.[173] According to Whitcomb and Morris:

"The gross tonnage of the Ark (which is a measurement of cubic space rather than weight, one ton in this case being equivalent to 100 cubic feet of usable storage space) was about 13,960 tons, which would place it well within the category of large metal ocean-going vessels today."[174]

And the number of creatures needed to arrive at our present-day biological construct is, again, according to Whitcomb and Morris:

"For all practical purposes, one could say that, at the outside, there was need for no more than 35,000 individual vertebrate animals on the Ark."[175]

---

[171] John C. Whitcomb and Henry M. Morris, The Genesis Flood/The Biblical Record and its Scientific Implications. Grand Rapids; Baker Book House, 1961, p. 10.
[172] John C. Whitcomb and Henry M. Morris, The Genesis Flood/The Biblical Record and its Scientific Implications. Grand Rapids; Baker Book House, 1961, p. 10.
[173] Holy Bible, King James Version, in public domain, Genesis 6:16.
[174] John C. Whitcomb and Henry M. Morris, The Genesis Flood/The Biblical Record and its Scientific Implications. Grand Rapids; Baker Book House, 1961, p. 10.
[175] John C. Whitcomb and Henry M. Morris, The Genesis Flood/The Biblical Record and its Scientific Implications. Grand Rapids; Baker Book House, 1961, p. 69.

The Ark had a capacity of approximately 125,280 animals, based on the calculations of size of and the fact that it could carry 522 stock cars.[176]

" ...at least 240 animals of the size of a sheep could be accommodated in a standard two-decked stock car."[177]

Thus it was possible to fit enough animals into the Ark, their food and water supplies, and eight people.

What's more, the location of the Ark is found either at Mt. Ararat, as the Bible says in Genesis 8:4, or at a location in Turkey with archaeological findings by Ron Wyatt, which is also in the same region. The latter is now a National Park, but there is also a large man-made structure atop Ararat that many believe is the actual Ark. If you Google "Ararat Anomaly" you will find the images of the actual site on Mt. Ararat and of the Ark. In any event, it requires faith to believe and not actual proof. I do not have the authorization to print any of the pictures, but they can be found at these two following websites: https://en.wikipedia.org/wiki/Ararat_anomaly and http://wyattmuseum.com/discovering/noahs-ark.

FOSSIL GRAVEYARDS

If you Google Images the words "bone bed," "fossil graveyards," or even "fossils" you will see hundreds of images that show huge bone beds where many different types of animals are all fossilized together. You would not have any of these areas if there was a gradual change

---

[176] John C. Whitcomb and Henry M. Morris, The Genesis Flood/The Biblical Record and its Scientific Implications. Grand Rapids; Baker Book House, 1961, p. 69.

[177] John C. Whitcomb and Henry M. Morris, The Genesis Flood/The Biblical Record and its Scientific Implications. Grand Rapids; Baker Book House, 1961, p. 69.

in the Earth. When animals died, they would be eaten or decayed and there would be no fossils at all.

You must have a rapid burial in order to get fossils

in any construct. This is an indication of why a catastrophic flood construct needs to be considered. Particularly, the Nebraska Bone Bed in Agate Springs is one that you will want to check out.

Photo from Colorado Museum of Natural History Used by permission from The Deluge Story in Stone by Byron C. Nelson, 1968, Bethany House Publishers, Minneapolis, MN 55438.

In the book <u>The Deluge Story in Stone</u> by Byron C. Nelson, many examples are cited; as many as 9,000 complete animals are buried together in one hill.[178] This type of occurrence might only be explained by a catastrophic flood module.

This is a picture I took at the Smithsonian in Washington. Fossils require a rapid burial for them to exist. This fossil would not exist unless it had been rapidly buried as in the Great Flood. This could be an indication that man and dinosaurs lived at the same time, although not necessarily in the same place if indeed the Great Flood is how they were buried and fossilized. [179]

---

[178] Byron C. Nelson, Th.M., <u>The Deluge Story in Stone.</u> Minneapolis, Minnesota. Bethany Fellowship, Inc. Publisher, 1968, p. 99
[179] Owned image.

Photo from Smithsonian Institute

Used by permission from The Deluge Story in Stone by Byron C. Nelson, 1968, Bethany House Publishers, Minneapolis, MN 55438. [180]

There are fossil crinoids in Iowa and elsewhere; such creatures only live now at ocean depths of 600 feet to a mile.[181] How could these exist in Iowa and elsewhere? Also, there are ripple marks that became fossilized—an indication of rapid sedimentation, which can only be explained by flood conditions.[182]

---

[180] Byron C. Nelson, Th.M., <u>The Deluge Story in Stone.</u> Minneapolis, Minnesota. Bethany Fellowship, Inc. Publisher, 1968, p.101.
[181] Byron C. Nelson, Th.M., <u>The Deluge Story in Stone.</u> Minneapolis, Minnesota. Bethany Fellowship, Inc. Publisher, 1968, p.101.
[182] Byron C. Nelson, Th.M., <u>The Deluge Story in Stone.</u> Minneapolis, Minnesota. Bethany Fellowship, Inc. Publisher, 1968, p.101.

Here is one example of perfectly preserved ripple marks that indicate a very rapid burial of all sizes in sedimentary deposits. This is indeed an indication of a catastrophic flood rather than uniformitarianism as a construct for our geological formations.[183]

Other important evidence supporting a universal flood is the fact that numerous upright petrified trees are distributed in the strata in the Bay of Fundy on the east coast of Nova Scotia.[184] This is an indication that the trees were buried spontaneously. This would be possible only in the catastrophic conditions of the flood.

Further, there is evidence for <u>transported,</u> rather than <u>in situ</u> origin of Nova Scotia coal.[185] While this evidence

---

[183] Byron C. Nelson, Th.M., <u>The Deluge Story in Stone.</u> Minneapolis, Minnesota. Bethany Fellowship, Inc. Publisher, 1968, p. 101.

[184] Harold G. Coffin, 'Research on the Classic Joggins Petrified Trees,' "Creation Research Society Annual," June 1969, pp. 35 -44.

[185] Harold G. Coffin, 'Research on the Classic Joggins Petrified Trees,' "Creation Research Society Annual," June 1969, pp. 35 -44.

by itself may not disprove evolution, it does at least raise the question as to why this information is not generally acknowledged in the examination of evolutionary theories.

THE PROBLEM OF MISPLACED FOSSILS

A radio program announcer proclaimed that one of the best "proofs" for evolution is the fact that, as he put it, *"You never find fossils that are of high evolutionary scale at lower strata!"*[186] This is just not the case. The geological column is pieced together from many different places around the world, and there are a large number of high-order and complex organisms being found in the lower strata, or even throughout the proposed geological column. Six of the major examples are recorded in the book The Creation Explanation by Robert E. Kofahl and Kelly L. Segraves. These include fossilized pollen grains being found in the lower levels of the Grand Canyon, even in Pre-Cambrian rocks.[187] European and Russian scientists have also reported numerous occurrences of fossil pollen grains.[188] *"But Pre-Cambrian rocks are dated as being older than 600 million years and were supposedly deposited hundreds of millions of years before the pollen-bearing plants evolved."*[189] There is much more proof that evolutionists have refused to deal with. *"Fossils of many kinds of woody stemmed plants have been discovered in the geological column in rocks assumed to represent*

---

[186] Clive Thomas, W.K.I.S. Radio Program, Feb. 1986.

[187] Robert E. Kofahl and Kelley L. Segraves, The Creation Explanation, Wheaton, IL, Harold Shaw Publisher, 1975, pp. 51 – 55.

[188] Robert E. Kofahl and Kelley L. Segraves, The Creation Explanation, Wheaton, IL, Harold Shaw Publisher, 1975, p. 53.

[189] Robert E. Kofahl and Kelley L. Segraves, The Creation Explanation, Wheaton, IL, Harold Shaw Publisher, 1975, p. 53.

periods predating by many millions of years the supposed time of their evolution. "[190]

HUMAN FOOTPRINTS IN CRETACEOUS STRATA

Finally, both human and dinosaur footprints have been found in the same level of strata (Cretaceous), demonstrating the contemporaneous existence of both man and dinosaur! This bears repeating again in this chapter because there are abundant examples found in the Paluxy River in Texas, discovered as long ago as 1939, and recorded in the magazine *Natural History* by Roland T. Bird in the article, "Thunder in His Footsteps," pp. 254 to 257. This evidence is simply ignored, called a "hoax" or, as described by Bird, as " *...some hitherto unknown dinosaur or reptile.* "[191]

There are numerous examples of this occurrence, as Whitcomb and Morris remark, *"But what appear to be human footprints have been found in rocks as early as the Carboniferous Period, supposedly some 250,000,000 years old."*[192]

Similar large human prints have been found in Arizona, near Mt. Whitney in California, near the White Sands in New Mexico, and other places throughout the world.[193]

---

[190]Robert E. Kofahl and Kelley L. Segraves, <u>The Creation Explanation,</u> Wheaton, IL, Harold Shaw Publisher, 1975, p. 53.

[191] Roland T. Bird, 'Thunder in His Footsteps,' "Natural History," May 1939, p. 257.

[192] John C. Whitcomb and Henry M. Morris, <u>The Genesis Flood/The Biblical Record and its Scientific Implications.</u> Grand Rapids; Baker Book House, 1961, p. 172.

[193] John Morris, <u>Tracking Those Incredible Dinosaurs and the People Who Knew Them,</u> CLP Publishers, 1980, pp. 91 -185.

John Morris reports in the book, <u>Tracking Those Incredible Dinosaurs and the People Who Knew Them,</u> that: *"Analysis of some prints owned by Dr. Clifford Burdick ... show 'mud-up-push,' pronounced instep, a slight change in crystallization roughly following the toe and ball of the foot, etc."[194]*

Many researchers may contend that those prints are fake, but it is not proven. An article by Albert C. Ingalls, entitled "The Carboniferous Mystery" in *Scientific American*, vol. 162 of January 1940, p. 14 states that human-like prints are exceedingly numerous:

"On sites reaching from Virginia and Pennsylvania, through Kentucky, Illinois, Missouri and westward toward the Rocky Mountains ... and from 5 to 10 inches long have been found on the surface of exposed rocks, and more and more keep turning up as the years go by."[195]

Yet many paleontologists and others dismiss all evidence of the coexistence of man with dinosaurs, most likely to save face. As Whitcomb and Morris remark, *"These prints give every indication of having been made by human feet, at a time when the rocks were soft mud."[196]*

But Ingles says:

"If man or even his ape ancestor, or even that ape's ancestor's early mammalian ancestor, existed as far back

---

[194] John Morris, <u>Tracking Those Incredible Dinosaurs and the People Who Knew Them,</u> CLP Publishers, 1980, p. 117.

[195] Whitcomb and Morris quoting Albert C. Ingalls, 'The Carboniferous Mystery,' Scientific American," vol. 162, Jan. 1940, p.14.

[196] John C. Whitcomb and Henry M. Morris, <u>The Genesis Flood/The Biblical Record and its Scientific Implications.</u> Grand Rapids; Baker Book House, 1961, p. 173.

as in the Carboniferous Period in any shape, then the whole science of geology is so completely wrong that all geologists will resign their jobs and take up truck driving. Hence for the present at least, science rejects the attractive explanation that man made these mysterious prints in the mud of the Carboniferous Period with his feet. "[197]

Also discovered was a sandal print near Antelope Springs, Utah with several trilobite fossils embedded in it. The trilobite is the major index fossil used to identify Cambrian Strata, which can be dated according to the geological column as old as 500,000,000 years.[198]

Further, it is reported that at the Mining Academy in Freiberg in Saxony, there is a puzzling human skull composed of brown coal and manganiferous and phosphatic limonite. The coal is presumably Tertiary in age, possibly making it between 12 and 70 million years old, and co-existent with the rise of mammals and the development of higher plants.[199]

Additionally, numerous "living fossils" have also been discovered. These include the well-known Coelacanth, which was thought to be extinct for some 60 million years; and the Tuatara, the sole survivor of the reptilian order of the beak-heads, which should have been extinct for some 135 million years according to the standard evolutionary time scale—but which are living today.[200]

---

[197] Whitcomb and Morris quoting Albert C. Ingalls, "The Carboniferous Mystery," Vol. 162, *Scientific American,* " January 1940 p. 14.
[198] Robert E. Kofahl and Kelley L. Segraves, The Creation Explanation, Wheaton, IL, Harold Shaw Publisher, 1975, pp. 51 -55.
[199] Whitcomb and Morris quoting Albert C. Ingalls, "The Carboniferous Mystery," Vol. 162, *Scientific American,* " January 1940 p. 14.
[200] John C. Whitcomb and Henry M. Morris, The Genesis Flood/The Biblical Record and its Scientific Implications. Grand Rapids; Baker Book House, 1961, p. 117

Finally, an article by Albert C. Ingalls from *Scientific American*, Volume 162, January 1940, page 14, entitled "The Carboniferous Mystery," gives considerable evidence that man and dinosaurs coexisted due to the finding of apparent "human" footprints in Carboniferous Strata at numerous sites. These prints are discounted and written off as non-human in nearly all cases, solely because they do not fit well with the prescribed evolutionary theory. Ingalls states:

"If man made these prints in this manner, than man's antiquity is no matter of a mere million years or so, as scientists think, but a quarter of a billion years, for they are found in rocks of the Carboniferous Period and those rocks were laid down about 250,000,000 years ago."[201]

Ingalls continues:

"Confronted with this claim, the scientist exclaims, 'What? You want <u>man</u> in the <u>Carboniferous?</u> Entirely and absolutely – totally and completely – impossible. We admit we don't know what made the prints, but we do know one agency that didn't, and that is man in the Carboniferous."[202]

Once again, evolutionists will hold true empirical scientific observation in contempt, rather than accept the notion that God is responsible for our creation instead of evolution. If evolution were indeed the case, then these types of prints should not exist.

----

[201] Whitcomb and Morris quoting Albert C. Ingalls, "The Carboniferous Mystery," Vol. 162, *Scientific American,*" January 1940 p. 14.
[202] Whitcomb and Morris quoting Albert C. Ingalls, "The Carboniferous Mystery," Vol. 162, *Scientific American,*" January 1940 p. 14.

A far more extensive study could be conducted which would show sufficient evidence for the catastrophic flood recorded in the book of Genesis, but the evidence listed here should be sufficient for anyone with an open mind. The excellent book, <u>The Genesis Flood / The Biblical Record and its Scientific Implications,</u> by John C. Whitcomb and Henry M. Morris, is indeed an extensive endeavor, worthy of note and further study. Nearly every ancient culture and society have a flood account. In Luke 17, verses 26 & 27, Jesus states:

"And as it was in the days of Noah, so shall it be in the days of the Son of Man. They did eat, and drank, they married wives, they were given in marriage, until the day that Noah entered into the ark, and the flood came, and destroyed them all."[203]

Aside from the flood, evolution is not supported by geological formations. Evolution is a fallacy unsupported by any hard evidence in the field of scientific geology.

This concludes the bulk of what I wrote in my previous book <u>Finding Proof of Jesus</u> concerning the Flood.

From this chapter, we can conclude that since there was a Worldwide Global Catastrophic Flood that occurred according to biblical texts, the Bible is shown to be true which accounts for our current biological construct of creatures alive today and there was no evolution of matter coming into existence from nothing. The Earth and the Universe are relatively young, therefore, there are no aliens that would have had time to evolve. They are, therefore, demons and not aliens that are piloting the UFOs we see today.

---

[203] Holy Bible King James Version, in public domain, Luke 17: 26 & 27.

Chapter 19
MEN IN BLACK

There are men that actually go around intimidating people who see a UFO and they talk about it! If it were a movie, that would be one thing, but this is real. In the movie *Men in Black*,[204] it was a sci-fi-comedy that showed these guys running around stopping the aliens from taking over the Earth, but in reality, the real Men in Black are actually helping the aliens to take over the Earth.

They are the enforcers of the plot of the Demonic Alien Overlords to ensure they are kept secret until they strike! People who knew too much and revealed what they knew have mysteriously died, and sometimes their deaths are ruled to be suicides when the person was definitely not suicidal. According to one History Channel YouTube video, T. Allen Greenfield alleges that author Dr. Morris K. Jessup who wrote the book <u>The Case for the UFO</u> was allegedly killed by the Men in Black in a faked suicide. Also, Frank Edwards died of an apparent heart attack, even though he was generally in good health. He was a radio broadcaster who wrote <u>Flying Saucers-Serious Business</u> and Jim Keith who wrote <u>Casebook of the Men in Black</u> also died of a sprained ankle.[205] None of them were over 50 years old, and all in good health.[206]

The cost of revealing what you know about UFOs is often death! If you work at Area 51, you are not able to

---

[204] *Men in Black,* American movie produced by Walter F. Parkes and Lorie MacDonald. 1997

[205] *UFO Hunters: Terrifying Encounters with Mysterious Beings (S3, E12) | Full Episode | History*
https://www.youtube.com/watch?v=w7z7u6enuNw&ab_channel=HISTORY Accessed 22.7. 2021

[206] Ibid.

reveal what you are working on. The security that keeps people out of Area 51, although perhaps a private security firm, until recently had a sign up that said: "Deadly Force Authorized". You can visit the area in Nevada today, but anytime people may have strayed onto the site, past the posted sign, they may have met with an untimely death. There are countless YouTube videos and articles about the Camo-dudes of the security forces that watch people that get too close to the sign and are ready to pounce on people that cross the line. There are most often two people in an unmarked car that will show up at your home or business and intimidate people into being silent that talk about what they know about UFOs in public, providing the person has insider knowledge. They are, to my knowledge, always dressed in black suits with a hat.

They have, according to witnesses that have seen them, have unusual features and may sometimes have robotic-type motions.[207] The question then becomes, are they normal men or perhaps even human-demon-alien-hybrids? I say alien because a demon in human form would be quite alien, although not an alien that is actually from another planet. It has long been rumored that the humanoid-demon bodies are being grown at Area 51, and this would be what the United States Government Deep State is working on with them, trading our genetic material, human and animal, with them for technology! I, of course, have no proof of this, but the Bible says it will be like it was back in the days of Noah, when genetic manipulation by angels mating with humans was happening before. I am going to rewrite the Scripture here again because it is so important.

---

[207] Ibid.

<blockquote>

But as the days of Noah were, so shall also the coming of the Son of man be. Matthew 24:37

And also:

There were giants in the earth in those days; and also after that, when the sons of God came in unto the daughters of men, and they bare children to them, the same became mighty men which were of old, men of renown. Genesis 6:4

</blockquote>

It is at least, hinted at here, that the genetic manipulation that was going on then, will continue even now. So, these Men in Black, that appear to be men, may in fact be the same type of genetically manipulated hybrids that were around at the time when the world was destroyed by the Flood to get rid of them. Where did we find men willing to intimidate people into keeping quiet about UFOs? Even to kill people when they "know too much"?

It started even back at the time of the Roswell Crash. People that saw the wreckage or the bodies were intimidated into keeping quiet about what they saw or knew. Back then, it was military types that visited the people. They made people that had any artifacts or pieces of the wreckage give them to the government. They spent days gathering up every piece of wreckage from the site. You won't find any evidence there. Today, they have a special group of men that will intimidate you if you reveal their agenda, the Men in Black!

If you ever watched the movie about the crash, you wonder how they could keep this secret all these years. The Kennedy assassination, by more than one gunman, was kept secret all these years, by the government covering everything up. They can do that, intimidation and cover-up! Cover up and intimidate! Provide false information, mixed with some truth, and pretty soon you won't believe your eyes. You will just accept what they say because you can't prove otherwise. If you try to say something the government doesn't like, like maybe that the 2020 election was rigged, look out, the thought police will be on your doorstep tonight. Even though there is ample evidence of fraud, you have to shut up! Freedom of Speech is no longer a part of our society! <u>1984</u> only missed it by a few years! In writing this book, the publication is doubtful, but unless we fight lies with truth, we are doomed!

"I am not suicidal and if anything happens, I didn't kill myself!"

Just thought I would put that in there just in case! You never know what might happen after this book comes out. I haven't seen the bodies of the so-called aliens that crashed at Roswell. I haven't seen any demon-human-hybrid beings as yet, that I know of, but I can surmise that since it happened back in the Days of Noah, it is happening still now, and being kept secret this time until the devil is ready to reveal his plan. There are people that would unwittingly help these demons do what they want because they will have believed the lie that they are from outer space. Magic tricks fool some people into thinking the magician has special powers, but in reality, it is smoke and mirrors. Likewise, these creatures are misleading people into thinking they are aliens, but in fact are demons. The Men in Black also are another way we can tell they are demonic. Why would an alien race, much more technologically advanced than we are, here to help us with all our problems, need to keep themselves secret from the world? They wouldn't. They would have shown themselves to us when they first arrived. Demons, however, don't reveal themselves to us, but hide and try to convince us they don't exist! So, these Men in Black are keeping the secret at all costs, even killing those that get in their way.

What can we determine about the Men in Black? Without some concrete evidence for them, like an admission by Congress or one of them, one YouTube video does not a conspiracy make. Except now that the United States Government has admitted that UFOs are real, there needs to be a pilot operating them in most cases, and that creature would be either an alien or a demon, or both. My money is on the demons because we know they exist without any question. Since God is real, Satan is real, and his objective from the beginning was to be God, therefore he will do whatever it takes to accomplish his goal.

Chapter 20
WHAT ARE UFOS?

The short answer is we don't know. If the government knows, they have covered it up. Here I am going off the deep end with no proof of anything, only to relay what I have heard over the years. Back in the Clinton administration, a man that had worked at Area 51 was dying of an unknown disease and the government would not reveal what chemicals were used so he could be treated. The man died, and they stated it was a National Security issue and could not be revealed. *The government let the man die, rather than to reveal what metal was being made!* Were it you or I that had this dilemma, they would let us die too!

So, what we do know is that the materials were of "unearthly" origin. Back when the Roswell Crash happened, and materials were brought into the police chief, they did not recognize it as simple weather balloon debris, but unknown materials. Then when Major Jessie Marcel investigated, he did not recognize them either as simple weather balloon debris. They later made him state it was weather balloon debris even after he showed it to his family, and they also knew it was something else. There was a coverup of gigantic proportions because, for several days after the crash, they kept bringing in the material from the crash site and shipped it to a base in Dayton, Ohio. The debris has been classified ever since. We most likely will never know what these materials are. *The fact that it is unknown material does not mean it is from another world, just that we don't know what the materials are.* Supposedly the materials could not be cut, drilled, or even when a torch was used, they could not make a dent. Further, one item, when folded or crumpled, it came right back to its original shape. *Demonic Entities could also know how to produce these materials, not only "aliens" from another planet.* I am going out on a limb without proof of what these beings are and say they are demons and not aliens. Again, to fulfill their agenda they will say or do anything to make us think they are something they are not.

Because our best minds can't find out what something is made of is not an indication it is alien or from another planet. I am reminded of a company that made soap. NASA came to the company president, I think Shaklee soap, and wanted to know the trade secrets for the soap, because it tended to make "water wetter." They said they needed to know because it was going on the Space Shuttle. He refused to tell them what was in it, and they were unable to determine what was in it. They never found out and still let it go on the Space Shuttle. We are not "all-knowing" when it comes to what things are made of. Likewise, because we are unable to determine what is the composition of certain metals, it does not mean it is from outer space. Demons are from here and could develop strange crafts that use materials from the Earth. We may just be unable to tell what they are made of. Now if we can't even tell what the chemical composition of soap is, we are obviously ignorant of a lot of things. But, getting back to what the UFOs are actually made of, we could not know, and we are in the dark about what they are constructed of.

HOW DO THEY OPERATE?

Since Chuck Yeager broke the sound barrier back in 1948[208] with a huge sonic boom the speeds that these

---

[208] From Wikipedia: https://en.wikipedia.org/wiki/Chuck_Yeager Accessed 8.11.2021

UFOs travel has been in contention. They make no sonic boom when they break the sound barrier and make extreme maneuvers that conventional aircraft can't. The consensus is that the maneuvers they make are far beyond what conventional aircraft can do, and when they are seen operating, they perform unexplainable maneuvers that we can't explain. They make turns at 90 degrees and accelerate to amazing speeds in an instant. The explanation is they are traveling in a "bubble", and this would be the reason they can operate beyond the scope of what we can in our vehicles. The fact that a crash happened at Roswell, however, shows they are not infallible. To get the tremendous speeds, they use an element called 115 that according to Bob Lazar,[209] is an antimatter drive to produce the huge speeds and acceleration that is observed. It decays into element 116[210] which is unstable and produces the antigravity that is accomplished. The element 115 (for us an unstable element) is placed inside a chamber (tube) and is bombarded with a proton and this decays into element 116 and a huge amount of energy is produced. This process produces antimatter which is the power source for the UFOs.

This is a very simple explanation of a complex process, for which there is obviously no proof. How they operate, providing they are real, would mean we are also aware that these crafts have occupants in them. This is all hearsay about how they may operate, and who the occupants may be. These occupants, I contend are not aliens,

---

[209] From a YouTube video: https://www.youtube.com/watch?v=DoI5QbNKVsY&ab_channel=JakeBroe Accessed 8.11.2021

[210] From a You Tube video: https://www.youtube.com/watch?v=VmJLSuLmgdg Accessed 8.11.2021

but demons. I have no proof of anything I am saying, either. Aside from the fact that no mention is made in the Bible about other worlds or aliens, but demons are described in detail, these creatures must be demons, who have developed craft to transport their humanoid bodies around in. Because we don't understand their technology, or their origin, does not prove they are from another planet. One thing about Satan and any demons, they are liars and would try to convince us of their "alienness" rather than being demons. When the Bible talks about angels, it tells us we are just below them, as I will again quote here:

These creatures, therefore, would have to be angels

> **For thou hast made him a little lower than the angels, and hast crowned him with glory and honour.** Psalm 8:5

that fell with the devil when they rebelled against God. Hence, ***they are not aliens, but they are demons***. I have yet no proof from the standpoint of high knowledge, but when we see how they act, that is how demons do and would act. They are offering their technology to us, again in exchange for DNA materials, and we are gullible enough to accept such a bargain. *If aliens, why not just reveal themselves to us? That would make sense. Why hide? If demons they would hide behind their ability to lie and make us think they are aliens, come down to little old Earth just to help us!* All the science fiction movies I have seen, depict them as what they would be, evil and not benevolent. They would want to exploit what the Earth has, both in people, commodities, and life, and move on. They would not be nice! I again have no proof of any of this, but if all things are equal because they act like demons, they must be demons. They are waiting for the right time to finally strike, just as the Rapture happens, and claim they are responsible for what has happened to the people remaining. Right now the Holy Spirit is restricting what they can do to us. After the Rapture, the Holy Spirit is removed from the Earth. This verse suggests that:

> **In a moment, in the twinkling of an eye, at the last trump: for the trumpet shall sound, and the dead shall be raised incorruptible, and we shall be changed.** I Corintihinians 15:52
> And after this:
> [14] **And deceiveth them that dwell on the earth by the means of those miracles which he had power to do in the sight of the beast; saying to them that dwell on the earth, that they should make an image to the beast, which had the wound by a sword, and did live.** Revelation 13:14

Notice that after the Rapture, things go haywire and that is when the demons will strike! Now they are restricted, although we know they would like to lie to us and tell us they are from some faraway planet in some alternate universe, they have to wait until we will swallow it.

Although I can't find the quote, a high-level representative stated, there is no proof they are aliens from another planet. If they are aliens, from a distant planet then prove it! Since no proof is offered to demonstrate they are aliens, from this chapter we can conclude that the UFOs and their occupants are in fact demons and not aliens.

Chapter 21
ARE CHRISTIANS BEING ABDUCTED?

I contend that if you are a Christian, you will never be abducted, neither can you be. I have no proof of this either. I can tell you from my experience back when I was oppressed by the demonic spirit, that it was not able to possess me. Likewise, since these creatures are demons, they will not abduct Christians. They will abduct those who have no protection from the Holy Spirit. Where does the Bible say this? To my knowledge, again this is not stated about aliens being able to abduct us, but there are indications we are at "war" against these Demonic Alien Overlords. I give you this Scripture:

> **[10] Finally, my brethren, be strong in the Lord, and in the power of his might.**
>
> **[11] Put on the whole armour of God, that ye may be able to stand against the wiles of the devil.**
>
> **[12] For we wrestle not against flesh and blood, but against principalities, against powers, against the rulers of the darkness of this world, against spiritual wickedness in high places.**
>
> **[13] Wherefore take unto you the whole armour of God, that ye may be able to withstand in the evil day, and having done all, to stand.**
>
> **[14] Stand therefore, having your loins girt about with truth, and having on the breastplate of righteousness;**
>
> **[15] And your feet shod with the preparation of the gospel of peace;**
>
> **[16] Above all, taking the shield of faith, wherewith ye shall be able to quench all the fiery darts of the wicked.**
>
> **[17] And take the helmet of salvation, and the sword of the Spirit, which is the word of God:**
>
> **[18] Praying always with all prayer and supplication in the Spirit, and watching thereunto with all perseverance and supplication for all saints;** Ephesians 6: 10-18

Notice how the spiritual wickedness is in "high places" in verse 12 that would mean they would be above us. We might also think they were ruling over us like politicians that are in charge. Today, politicians seem more and more controlled by evil than ever before. An example is an abortion. Today it is legal to kill a baby even after it is born in some states. This will keep the child from becoming a Christian. Kill it before it is born, even after, to ensure it will not become a Christian and be kept safe from being able to be abducted. In Virginia, the governor declared that if a child was unwanted or survived an abortion attempt, the staff should keep it comfortable until it died! What kind of evil is this? This can only be coming from demonic control, even though the people will deny it. Who knows really if the person in the office made some type of "pact" with the devil to put them in power, and then once in power, they do the bidding of the devil? All politicians must be secretly controlled by some form of demonic influence.

In the United States annually, according to Wikipedia, 862,300 abortions were performed in 2017. That number is increasing every year. Left unchecked we would just kill babies before they are born. We can have sex and not have any consequences for our actions. If a woman gets pregnant, just kill the child. When did that become normal? It fulfills prophecy to show how and what is happening. Consider this Scripture:

> **And because iniquity shall abound, the love of many shall wax cold.** Matthew 24:12

Women and men no longer care for their children in our society today. We have sex and "throwaway" the result. This used to not be the case, but today people are being influenced by demons more than ever. *Perhaps the craft they are using to fly over us at night somehow contributes to the sinful influences we have allowing them to control more and more of our lives.* Tactics of demons are pervasive, and today we are too sophisticated to say that what compels us are *waves of sin,* somehow being broadcast out from craft being piloted by Demonic Alien Overlords compelling us to sin in ways that man could not be responsible on his own. Child molestation, pornography, child abductions, murder, rape, you name it, are controlling our society today like never before. I contend, again without proof, that the sophistication of the devil in convincing us to sin, is designed just to make us do what the devil wants. He always was able to get us to do wrong before, but now, his actions are seemingly much more concise in their abilities. Before, in the beginning, the devil lied to Eve and got her to sin. We all know that story. *Now I am suggesting that the devils' tactics have changed, in by flying in craft above us, they can shoot out waves of sin upon our brains to cause us to do what they want.* Do you have a better explanation of how the devil does it? I know that for me, it is when I am trying to sleep, or when I get tired, the devil comes at me and tries to get me to sin. He also tries at times when I am alone when I think no one will know, that he tries to convince me to sin. You may have a different way the devil provokes you. But exactly *how* does the devil make us want to sin? You say, well, it enters my mind, and then I want to do it. But *how exactly does it enter your mind?* The Bible does not say specifically but gives this explanation:

> **But every man is tempted, when he is drawn away of his own lust, and enticed.** James 1:14

But how are we enticed? Maybe it is just this simple: Demonic Alien Overlords ride around in crafts above our heads and shoot down at us the *waves of sin* in which we will be most affected by and that then *entices* us to commit those sins.

I will give you an example of one time when I could have sinned but did not. I went to a store and bought items and when I got to my car, I realized I didn't pay for an item. I could have overlooked it and just went home. I went back into the store and told them. I bet it never happened before. I paid for the item and left. Other people are affected by what they can steal from companies, or what they can get away with. You see people looting stores any time they get the chance. There is now in many cities no penalty. I am just not that way. I have other sins that I will do, and so do you. So, we are "targeted" by the devil in what types of sins he can get us to do, and perhaps he shoots down the "targeted" waves of sin on to each person he can capture with his "targeted" *waves of sin,* shot down from the craft the demons are flying around in over our heads.

This is very far-fetched, but not really. You don't know how the devil tempts us and neither do I, we just know we are tempted. And most times we will do what we are tempted to do. We have trouble resisting, if we didn't, we would not need to go to church, and maybe we avoid church because we are sorry we did what we did. However, we are tempted, whether, by a traditional method of the devil, or now by more sophisticated methods, we still sin. But the devil is limited in his ability, regardless of how he does it. It tells us that we are protected as Christians from the devil. I give you these Scriptures:

> **No weapon that is formed against thee shall prosper; and every tongue that shall rise against thee in judgment thou shalt condemn. This is the heritage of the servants of the LORD, and their righteousness is of me, saith the LORD.** Isaiah 54:17
>
> And also:
>
> **Being confident of this very thing, that he which hath begun a good work in you will perform it until the day of Jesus Christ:** Philippians 1:6

We are protected from the devil, unlike those that are not Christians, and therefore, we can't be abducted. We most likely won't even see a UFO because they have no reason to reveal themselves to us, but we still might. Every time I thought I saw something, there was always an explanation. One time it rained hard, and my brother saw a light in a field. It slowly faded away. It was a reflection of light, maybe an outside light in a puddle, and when the wind blew it disappeared. Another time, when flying in a private plane, we all saw a huge crescent shape moving away from us at blinding speed. That was a UFO for sure, right? But as we flew we passed over a huge building that had the sun reflecting off the roof. It wasn't a UFO at all, just an optical illusion. Even with the admission by the government that UFOs are real, we may still not see one, and not only so, but we will also not be abducted. Our protection by God extends beyond what the devil can do to us or with us. Now, can I claim that no Christians have ever been abducted, or that none are being abducted now? No, absolutely not! I am not able to determine who is a Christian or not. I am saying it is less likely that it would be Christians who are abducted, rather than non-Christians. In Christian circles, I have no knowledge of any Christian saying they were abducted after they became a Christian. The abductions seem to be pervasive in non-Christian circles. We can conclude the following about what we know about demons: ***TO AVOID BEING ABDUCTED AND MOLESTED BY DEMONIC ALIEN OVERLORDS, BECOME A CHRISTIAN!*** In the end, after the Rapture, it will become very difficult to make this decision. Make it now before it is too late! Refer back to Chapter 5 on how to be saved and make the choice now! Otherwise, be prepared to go through Hell on Earth after the Rapture, and be forced to take the Mark of the Beast which will be an implant, that once you take it, you will be powerless to resist what the devil wants you to do!

Chapter 22

HOW DO WE DEFEAT THE DEMONIC ALIENS?

Obviously, you have no chance if you are not a Christian! To say that we can stand against the devil on our own strength is ludicrous! I, even as a Christian, am a complete failure when it comes to overcoming those sins, I am susceptible too and surely, you are too. That is why we absolutely need Jesus. If we could keep from sin, and just live our lives without sin, we would not need a Savior, and the sacrifice Jesus made is of no effect. Satan would be ruling us in ways we can't imagine, even worse than now. But there is a way for us to defeat these Demonic Alien Overlords by, first of all, praying! How simple is that? Pray and ask God to help us. Simply put, the Scripture says simply this:

> **Pray without ceasing**. I Thessalonians 5:17

Obviously, if you are doing this, you would not be sinning! But this unfortunately only applies to Christians! You first, obviously must be a Christian, or none of this means anything. (Refer to How to Become a Christian in chapter 5). I only covered the tip of the iceberg in that chapter about how we defeat them. In context, the Bible gives in I Thessalonians 5, a more detailed explanation of what we need to be doing. The whole chapter needs to be reviewed about how to defeat them.

[1] But of the times and the seasons, brethren, ye have no need that I write unto you.

[2] For yourselves know perfectly that the day of the Lord so cometh as a thief in the night.

[3] For when they shall say, Peace and safety; then sudden destruction cometh upon them, as travail upon a woman with child; and they shall not escape.

[4] But ye, brethren, are not in darkness, that that day should overtake you as a thief.

[5] Ye are all the children of light, and the children of the day: we are not of the night, nor of darkness.

[6] Therefore let us not sleep, as do others; but let us watch and be sober.

[7] For they that sleep in the night; and they that be drunken are drunken in the night.

[8] But let us, who are of the day, be sober, putting on the breastplate of faith and love; and for an helmet, the hope of salvation.

[9] For God hath not appointed us to wrath, but to obtain salvation by our Lord Jesus Christ,

[10] Who died for us, that, whether we wake or sleep, we should live together with him.

[11] Wherefore comfort yourselves together, and edify one another, even as also ye do.

[12] And we beseech you, brethren, to know them which labour among you, and are over you in the Lord, and admonish you;

[13] And to esteem them very highly in love for their work's sake. And be at peace among yourselves.

[14] Now we exhort you, brethren, warn them that are unruly, comfort the feebleminded, support the weak, be patient toward all men.

[15] See that none render evil for evil unto any man; but ever follow that which is good, both among yourselves, and to all men.

[16] Rejoice evermore.

[17] Pray without ceasing.

[18] In every thing give thanks: for this is the will of God in Christ Jesus concerning you.

[19] Quench not the Spirit.

> ²⁰ **Despise not prophesyings.**
> ²¹ **Prove all things; hold fast that which is good.**
> ²² **Abstain from all appearance of evil.**
> ²³ **And the very God of peace sanctify you wholly; and I pray God your whole spirit and soul and body be preserved blameless unto the coming of our Lord Jesus Christ.**
> ²⁴ **Faithful is he that calleth you, who also will do it.**
> ²⁵ **Brethren, pray for us.**
> ²⁶ **Greet all the brethren with an holy kiss.**
> ²⁷ **I charge you by the Lord that this epistle be read unto all the holy brethren.**
> ²⁸ **The grace of our Lord Jesus Christ be with you. Amen.** I Thessalonians 5

Further, other verses in the Bible tell us how to defeat them. Notably when the Bible says this:

> **And they overcame him by the blood of the Lamb, and by the word of their testimony; and they loved not their lives unto the death.** Revelation 12:11

Then, though we would have to die because, in the end, it is too late! Notice it says *death! You have to prove your resolve by dying!* Before you get to this point, make sure you are reading the Word of God *daily!* Start with a Bible that is through the Bible in one year and stick to it! For years I used *Bible Gateway,*[211] an online Bible because it reads it to you. You can read a Bible manually, or have it read to you, just put the Word of God in your heart! Stop doing these things for a moment and it allows an access point to the Demonic Alien Overlords. You may in fact disagree with the idea that these beings that pilot these crafts are demonic, but when you look at how they act and what they are doing, as stated in chapter 13, their actions indicate they are operating like demons, and to combat demons, we need a spiritual approach. I could

---

[211] https://www.biblegateway.com/ Accessed 13.8.2021

be totally wrong, but I and others have stated they are demonic. The "proof is in the pudding" as they say, and this pudding is tasting very bad! What other Scriptures teach us to combat the devil? Jesus, when he was tempted is the best example.

[1] And Jesus being full of the Holy Ghost returned from Jordan, and was led by the Spirit into the wilderness,
[2] Being forty days tempted of the devil. And in those days he did eat nothing: and when they were ended, he afterward hungered.
[3] And the devil said unto him, If thou be the Son of God, command this stone that it be made bread.
[4] And Jesus answered him, saying, It is written, That man shall not live by bread alone, but by every word of God.
[5] And the devil, taking him up into an high mountain, shewed unto him all the kingdoms of the world in a moment of time.
[6] And the devil said unto him, All this power will I give thee, and the glory of them: for that is delivered unto me; and to whomsoever I will I give it.
[7] If thou therefore wilt worship me, all shall be thine.
[8] And Jesus answered and said unto him, Get thee behind me, Satan: for it is written, Thou shalt worship the Lord thy God, and him only shalt thou serve.
[9] And he brought him to Jerusalem, and set him on a pinnacle of the temple, and said unto him, If thou be the Son of God, cast thyself down from hence:
[10] For it is written, He shall give his angels charge over thee, to keep thee:
[11] And in their hands they shall bear thee up, lest at any time thou dash thy foot against a stone.
[12] And Jesus answering said unto him, It is said, Thou shalt not tempt the Lord thy God.
[13] And when the devil had ended all the temptation, he departed from him for a season.
[14] And Jesus returned in the power of the Spirit into Galilee: and there went out a fame of him through all the region round about. Luke 4: 1 - 14

Notice, that Jesus didn't dispute that the devil owns the world. Since the Fall of Man, he is in charge of the world. He owns it all and is the ruler of this world. What Jesus did do is to use the Scripture to rebuke the devil from his temptations. These Demonic Alien Overlords don't want to reveal themselves as being demons, because there is a way to defeat them. They want to masquerade as aliens so we will be in "awe" of them and their technological powers. They will make people well of diseases, that they caused, and tell us they are good. Once we accept them as healed of cancer, diabetes, lame people will be made to walk, etc., all with the implant they take into their hand or forehead! Then it changes us into their slaves, and we have no way out! Mark my words, this is exactly what is about to happen, gradually at first, then full-blown disclosure! The Rapture happens, and then they are revealed as aliens from outer space! In any event, if they say they are from, a galaxy far, far away, how would we prove they're not? I say test them to see if they are who they say they are, and you will find them to be demons! They essentially win, because who can dispute the "good" they are doing? People get healed of cancer, diabetes, even lame people will be made to walk, even grow limbs, perhaps, all with the implant they take into their hand or forehead! Then it changes us into their slaves, and we have no way out! Be prepared for when they arrive as they land at the White House and they revealed as aliens from outer space! They will always lie and say they are from, planet X or Zeta reticuli, again how would we prove they're not? I say test them to see if they are who they say they are. Read the Bible and you will find them to be demons! I John 4 is a great example of how these creatures should "love" us! They do not display love, but death and lies and they hide from us!

¹ Beloved, believe not every spirit, but try the spirits whether they are of God: because many false prophets are gone out into the world.
² Hereby know ye the Spirit of God: Every spirit that confesseth that Jesus Christ is come in the flesh is of God:
³ And every spirit that confesseth not that Jesus Christ is come in the flesh is not of God: and this is that spirit of antichrist, whereof ye have heard that it should come; and even now already is it in the world.
⁴ Ye are of God, little children, and have overcome them: because greater is he that is in you, than he that is in the world.
⁵ They are of the world: therefore speak they of the world, and the world heareth them.
⁶ We are of God: he that knoweth God heareth us; he that is not of God heareth not us. Hereby know we the spirit of truth, and the spirit of error.
⁷ Beloved, let us love one another: for love is of God; and every one that loveth is born of God, and knoweth God.
⁸ He that loveth not knoweth not God; for God is love.
⁹ In this was manifested the love of God toward us, because that God sent his only begotten Son into the world, that we might live through him.
¹⁰ Herein is love, not that we loved God, but that he loved us, and sent his Son to be the propitiation for our sins.
¹¹ Beloved, if God so loved us, we ought also to love one another.
¹² No man hath seen God at any time. If we love one another, God dwelleth in us, and his love is perfected in us.
¹³ Hereby know we that we dwell in him, and he in us, because he hath given us of his Spirit.
¹⁴ And we have seen and do testify that the Father sent the Son to be the Saviour of the world.
¹⁵ Whosoever shall confess that Jesus is the Son of God, God dwelleth in him, and he in God.

> [16] And we have known and believed the love that God hath to us. God is love; and he that dwelleth in love dwelleth in God, and God in him.
> [17] Herein is our love made perfect, that we may have boldness in the day of judgment: because as he is, so are we in this world.
> [18] There is no fear in love; but perfect love casteth out fear: because fear hath torment. He that feareth is not made perfect in love.
> [19] We love him, because he first loved us.
> [20] If a man say, I love God, and hateth his brother, he is a liar: for he that loveth not his brother whom he hath seen, how can he love God whom he hath not seen?
> [21] And this commandment have we from him, That he who loveth God love his brother also. I John 4

Notice if someone (or something) is of God, it will be loving. These creatures are in no way loving! Be they aliens or Demonic Alien Overlords, they are not from God! Therefore, in this chapter, we can conclude that these creatures are demonic and not the aliens they masquerade as. You may still choose to believe they are aliens if you want to be naive, but I put my trust in a God that shows love, and these creatures do not show love. I need to add another reason why these creatures are not benevolent aliens to what I have already stated in the previous chapter:

True Aliens *are not* mentioned in the Bible, therefore do not likely exist.

We could not have evolved, therefore, neither could aliens.

Demons *are* mentioned in the Bible; therefore, any discovered so-called *aliens* are demons.

The lie they will state is that they "seeded" our planet, thus taking credit for God's creation.

The bases they have are here on Earth and not on some other planet, but they will lie and say they are from another planet. Government officials have stated there is no indication UFOs or UAPs are from another planet.

They are abducting people, doing medical experiments on us, collecting genetic material, and all manner of nefarious acts, even collecting sexual materials (sperm and eggs) therefore they are not benevolent aliens but are shown to be demons.

This collecting of genetic material would be for the purpose of making hybrid-alien creatures which is a continuation of what was done before in Genesis when the demons "mated" with human females and produced giants.

Angels are a little higher than us, thus fallen angels (demons) would be more technologically advanced than we are, as these creatures are.

When they are disclosed, people will worship them and forget about God, like the Bible says, the whole world will wonder after the beast.

They are implanting people with devices, which falls into place with the Bible where everyone will be required to get the mark to buy or sell, which the implant could in fact be a type of "mark."

Demonic entities control people and Demonic Alien Overlords flying above us would be in a position to do just that.

Other reasons demonstrating a young Universe.

They are sexually depraved.

They are not loving, therefore, are not of God!

Chapter 23
HITLER'S OBSESSION WITH ALIENS

I can no longer find the picture that showed Hitler meeting with a little gray alien, but it was on the Internet before. Although it was most likely a fake, it was true in many respects. During WWII Hitler received help from the Demonic Aliens and they helped him and his loyalist scientists to create some of the alien spacecraft types that we see in the skies today! They look just like the little "sport models" that were flying around in the sky back when this whole thing started. If he had obtained the keys to how to develop them, we most certainly would have lost the war. Here is an image of what they looked like:

212

As you can see from this picture, there is a striking resemblance between what we see flying around in the sky in many pictures and what we see in old pictures of many UFOs. The only thing they needed was the method of operating them, using element 115, and it would have been a disaster for us. How they helped him is in dispute, or how he made contact is the question. The story goes like this: He had psychics that helped him to contact

---

[212] Owned image from Shutterstock

them. Now I don't know how they did it, but supposedly they were women witches that contacted the demons and showed them the designs of their crafts. Whether he actually contacted them or not is again in dispute, but you can't deny the similarities here in the design. There were other designs as well besides the one in the picture. One such design was a "bell" design, which might have been like the crash that happened in Kecksburg, PA on December 9, 1965.[213] The Kecksburg UFO incident occurred on December 9, 1965, at Kecksburg, Pennsylvania, United States, when a fireball was reported by citizens of six U.S. states and Canada over Detroit, Michigan, and Windsor, Ontario.[214] This craft was later explained as a Soviet satellite that crashed, however, it changed directions according to reports.[215]

This craft was like the ones Hitler was also developing in design, and so if this was in fact an alien spacecraft of some type, Hitler was in some contact with these demons to develop this design, it would appear. The authorities closed the area and covered up the crash and the military took it out on a flatbed truck the night it crashed, as it was roughly the size of a VW.[216] None of this proves the involvement of Hitler in this design, but because it is similar, Hitler and his cohorts must have had insight into how these crafts and the one depicted above were designed. In the picture found with the article, there are very strange makings[217] which were on the craft that according to witnesses looked "otherworldly," I have heard. Regardless,

---

[213] From Wikipedia "Kecksburg UFO Incident," https://en.wikipedia.org/wiki/Kecksburg_UFO_incident Accessed 8.14.2021.
[214] Ibid.
[215] Ibid.
[216] Ibid.
[217] Ibid.

if alien, demonic, or conventional, the design is like some of the craft being developed by Hitler during the war. Although I don't have an image I can display here, I have seen similar designs from other documentaries that show Hitler developing this type of craft during the war. This one however fails in its comparison to the more elite design shown in the photograph. Whenever we think of flying machines, they would need to have control surfaces in order to operate, and these have none. Further, they would need a conventional propulsion system, which again is missing from these futuristic designs, therefore, it is apparent that Hitler was somehow in contact with these Demonic Aliens in order to develop these futuristic designs.

Hitler may have in fact been "possessed" by these Demonic Aliens! How could one man have such an oratory prowess to convince a whole nation to exterminate a people? He wound up killing over 8 million Jews in the ensuing turmoil of the war. Why would we go along with it? Today, for example, if a leader says we are going to round up all the Christians and put them in FEMA camps, would you go along? I can't mention any other group without being classified as racist, or worse, but this is exactly what will happen soon. They have these camps ready now just for this purpose, and all we need is for the government to implement the process! Once the Rapture happens, I tell you now, anyone who dissents from the worship of the Demonic Alien creatures will be arrested and locked away, and anyone who refuses to take the implant (Mark of the Beast) will be killed by guillotine.

Hitler had unusual supernatural powers beyond what we can fathom, and there is speculation that he indeed was possessed by demons. Perhaps even Satan himself. There is, of course, no proof of any of this, but the oratory powers he possessed over Germany, show something more than human. He was being "helped" by these Demonic Alien Overlords to kill the Jews and others, but why? The obvious answer was they are God's chosen people and killing them would be the goal of Satan. Since we can see that these creatures are now being revealed as real, they are going to control us once the Rapture occurs, I contend. It could be earlier, but everything points to the Rapture happening first and then the full disclosure.

Hitler, in developing these spacecraft, proves nothing as far as if the Demonic Aliens were helping him. If not, these designs were otherworldly in any case. Luckily, we won the war before he was able to get these crafts or the nuclear bombs we had developed. God helped us, or we would not have developed these weapons, I surmise. It is an indication at least that what he was doing was in line with the UFOs that were in the sky. There is no indication I have that shows these spacecraft landing or crashing in Germany as they did with us in Roswell, NM, but still, how did Hitler have the insight to develop these craft? I think we can conclude he most defiantly had help from them, just thank God it was too late for him.

Chapter 24
BIRTH PANGS OF THE RAPTURE

When will the Rapture occur? No one can say for sure, and if I do so I would be a heretic. I can say it is very near. It is the next great thing that will happen, and then the world will be without God's protection. But *when* will the Rapture occur? We do not have a specific date, but soon. With the disclosure of UFOs being real it is closer than we think. It describes the end in Luke 21: 7 – 36 as when we see pestilence. Covid-19 is indeed that! So, the end is near! The actual Demonic Alien Overlords will arrive and then we will be raptured out. It could happen anytime, simultaneously, or just after the Rapture! It could happen today!

[7] And they asked him, saying, Master, but when shall these things be? and what sign will there be when these things shall come to pass?

[8] And he said, Take heed that ye be not deceived: for many shall come in my name, saying, I am Christ; and the time draweth near: go ye not therefore after them.

[9] But when ye shall hear of wars and commotions, be not terrified: for these things must first come to pass; but the end is not by and by.

[10] Then said he unto them, Nation shall rise against nation, and kingdom against kingdom:

[11] And great earthquakes shall be in divers places, and famines, and pestilences; and fearful sights and great signs shall there be from Heaven.

[12] But before all these, they shall lay their hands on you, and persecute you, delivering you up to the synagogues, and into prisons, being brought before kings and rulers for my name's sake.

[13] And it shall turn to you for a testimony.

[14] Settle it therefore in your hearts, not to meditate before what ye shall answer:

[15] For I will give you a mouth and wisdom, which all your adversaries shall not be able to gainsay nor resist.

¹⁶ And ye shall be betrayed both by parents, and brethren, and kinsfolks, and friends; and some of you shall they cause to be put to death.

¹⁷ And ye shall be hated of all men for my name's sake.

¹⁸ But there shall not an hair of your head perish.

¹⁹ In your patience possess ye your souls.

²⁰ And when ye shall see Jerusalem compassed with armies, then know that the desolation thereof is nigh.

²¹ Then let them which are in Judaea flee to the mountains; and let them which are in the midst of it depart out; and let not them that are in the countries enter thereinto.

²² For these be the days of vengeance, that all things which are written may be fulfilled.

²³ But woe unto them that are with child, and to them that give suck, in those days! for there shall be great distress in the land, and wrath upon this people.

²⁴ And they shall fall by the edge of the sword, and shall be led away captive into all nations: and Jerusalem shall be trodden down of the Gentiles, until the times of the Gentiles be fulfilled.

²⁵ And there shall be signs in the sun, and in the moon, and in the stars; and upon the earth distress of nations, with perplexity; the sea and the waves roaring;

²⁶ Men's hearts failing them for fear, and for looking after those things which are coming on the earth: for the powers of Heaven shall be shaken.

²⁷ And then shall they see the Son of man coming in a cloud with power and great glory.

²⁸ And when these things begin to come to pass, then look up, and lift up your heads; for your redemption draweth nigh.

²⁹ And he spake to them a parable; Behold the fig tree, and all the trees;

³⁰ When they now shoot forth, ye see and know of your own selves that summer is now nigh at hand.

³¹ So likewise ye, when ye see these things come to pass, know ye that the kingdom of God is nigh at hand.

³² Verily I say unto you, This generation shall not pass away, till all be fulfilled.

³³ Heaven and earth shall pass away: but my words shall not pass away.

> **34** And take heed to yourselves, lest at any time your hearts be overcharged with surfeiting, and drunkenness, and cares of this life, and so that day come upon you unawares.
> **35** For as a snare shall it come on all them that dwell on the face of the whole earth.
> **36** Watch ye therefore, and pray always, that ye may be accounted worthy to escape all these things that shall come to pass, and to stand before the Son of man. Luke 21: 7 - 36

Jewish rabbi Yitzhak Kaduri saw a vision of Jesus. Rabbi Kaduri wrote the name of Jesus in a note shortly before his death in January 2006 and also proclaimed that he had seen the Messiah![218] He predicted that the Messiah would come very soon after the death of Ariel Sharon. Ariel Sharon died on January 11, 2014.[219] Each day that passes means we are ever closer to the return of Jesus, or the Rapture of the Church.

Once the Rapture happens, the Demonic Aliens will step in and proclaim they are our saviors. Jesus does not say this exactly, but what if it happened today that the spacecraft landed at the White House and out step the little gray alien creatures and proclaim they are here from an unknown planet? The world would begin to worship them. They would fill the void of the Jewish Messiah, whoever the creature is that is in charge, which I contend will be Satan himself! He would indeed be able to "call down fire from Heaven" like it says he can do in Revelation 13:13!

---

[218] Aviel Schneider, "The Rabbi, the Note and the Messiah," *Israel Today*, Thursday, May 30, 2013 (reprint of the story that first appeared in the April 2007 issue) http://www.israeltoday.co.il/NewsItem/tabid/178/nid/23877/Default.aspx, accessed Aug. 26, 2015.
[219] Wikipedia, "Ariel Sharon," https://en.wikipedia.org/wiki/Ariel_Sharon Accessed 8.15.2021

They would be hailed as highly advanced in technology, therefore, would fit the concept of God for many people. They will tell us to avoid ever getting sick and dying again all we need to do is take their implant they have for us, and we will not die! They would even permit the construction of the Jewish Temple! Who could fight against them? They would destroy all opposition. For the first three and a half years, everything will be great! Then things will fall apart. Whoever their ruler is he will enter the Temple and declare himself to be God! That will be the Antichrist, Satan himself, and we will have no choice but to go along with it or be killed.

I don't expect to be here then! Christians will have been raptured out; I hope. But if the Tribulation is something we do have to go through to prove we are Christians, we will be in dire straits! The three schools of thought are that the Rapture occurs Pre-Tribulation, Mid-Tribulation, or Post-Tribulation. I have the hope it is Pre-Tribulation. Otherwise, most so-called mediocre, namby-pamby Christians will not be able to stand unless the Rapture occurs before the Demonic Alien Overlords are disclosed, and the Tribulation time of seven years starts. We could still see them arrive and then the Rapture will happen, that might be ok, but if we stay for longer than a few weeks, we will fail under the weight of not being able to buy or sell or eat, because we will have no food unless we take the Mark of the Beast. I can't grow mine, and most likely neither do you. You can't possibly have enough food stored for you and your family for seven years, I wouldn't think! Even if you live on a farm and grow all your own food, you won't be able to keep it because it will be confiscated or stolen by people who are starving. It will be a very terrible time, beyond what any of us can imagine.

THE BIRTH PANGS

1. World War I – war and rumor of wars
2. World War II – war and rumor of wars
3. Concentration camps – war and rumors of war
4. Israel becomes a nation on May 14, 1948 – major birth pang
5. The Great Depression – economic turmoil
6. 911 – war and rumor of wars.
7. Covid-19 – pestilence
8. Different types of flues – pestilence
9. Continued terrorist attacks – war and rumors of wars
10. Recessions – economic turmoil
11. Mass shootings – wars and rumors of wars
12. Gas/energy shortages – economic turmoil
13. Storms and Earthquakes – weather turmoil
14. Global Warming/Climate Change fears -weather turmoil
15. Inflation – economic turmoil
16. Border crisis – People want out of communist countries and desperately try to enter the U.S.

These are only a few of our struggles that are happening in the 20th – 21st Centuries. It is now becoming a very fearful time. When was the last time you ever remember kids having to wear masks in school, or not being able to even attend classes because of a pandemic? When did the government last demand we all wear masks and get a vaccine to prevent the spread of any other disease? Men's hearts are failing them for fear, now and we are getting very close to the end!

> **Men's hearts failing them for fear, and for looking after those things which are coming on the earth: for the powers of heaven shall be shaken.** Luke 21:26

We need to prepare ourselves for the Return of Jesus now! We have to know it is close! We have to repent of our sins, or as the Scripture says, we may be left behind to suffer. If we are Christians we should be caught up in the air, but other verses state terrifying scenarios!

> **In a moment, in the twinkling of an eye, at the last trump: for the trumpet shall sound, and the dead shall be raised incorruptible, and we shall be changed.** I Corinthians 15:52

This is what we all hope for, and before the Tribulation starts. The word Rapture never occurs in the Bible, but this previous verse describes what is commonly known as the Rapture. But then it also states another scripture we need to head:

> [40] **Be ye therefore ready also: for the Son of man cometh at an hour when ye think not.**
> [41] **Then Peter said unto him, Lord, speakest thou this parable unto us, or even to all?**
> [42] **And the Lord said, Who then is that faithful and wise steward, whom his lord shall make ruler over his household, to give them their portion of meat in due season?**
> [43] **Blessed is that servant, whom his lord when he cometh shall find so doing.**
> [45] **But and if that servant say in his heart, My lord delayeth his coming; and shall begin to beat the menservants and maidens, and to eat and drink, and to be drunken;**
> [46] **The lord of that servant will come in a day when he looketh not for him, and at an hour when he is not aware, and will cut him in sunder, and will appoint him his portion with the unbelievers.**

> **⁴⁷ And that servant, which knew his lord's will, and prepared not himself, neither did according to his will, shall be beaten with many stripes.**
> **⁴⁸ But he that knew not, and did commit things worthy of stripes, shall be beaten with few stripes. For unto whomsoever much is given, of him shall be much required: and to whom men have committed much, of him they will ask the more.** Luke 12: 40 - 48

We all expect that if we are Christians we will not have to go through the Tribulation. But what if those who profess to be Christians, but are still practicing their same old sins are the ones who are "drunken?" We, as Christians, all have the Holy Spirit, of course, but we all are not functioning as "good" servants. I am not going to presume to say that some Christians will be raptured, and others will not, but we need to heed warnings like this one, or else we may find ourselves as part of the Tribulation, while other Christians were taken up. Just a warning, not an absolute truth.

THE MAJOR BIRTH PANG – ISRAEL BECOMES A NATION MAY 14, 1948

This happed right in line with prophecy, and exactly to the year![220] It is a complicated prophecy that shows how it happened right at the right time. It does not matter when, but that it did! The major prophecy is in Ezekiel and Jerimiah.[221] These are the major prophetic Scriptures

---

[220] **"Shalom from G-d – English"** Considering the Tanakh's promise of Shalom, the Man, and the Land – all in the sea of the Nations" https://en.shalomfromg-d.net/2017/05/17/the-hebrew-prophet-vs-jewish-skeptic-pt-2/?gclid=CjwKCAjw9uKIBhA8EiwAY-PUS3C_3jeRTzI2mZfeEl99hLzxpMEckbpRa-nYZdXL_339fbKVyaii3kWhoCVpsQAvD_BwE   Accessed 15.8.2021

[221] Ibid.

that tell us when it is about time for the Rapture to occur! Ezekiel chapter 37 says:

[1] The hand of the LORD was upon me, and carried me out in the spirit of the LORD, and set me down in the midst of the valley which was full of bones,

[2] And caused me to pass by them round about: and, behold, there were very many in the open valley; and, lo, they were very dry.

[3] And he said unto me, Son of man, can these bones live? And I answered, O Lord GOD, thou knowest.

[4] Again he said unto me, Prophesy upon these bones, and say unto them, O ye dry bones, hear the word of the LORD.

[5] Thus saith the Lord GOD unto these bones; Behold, I will cause breath to enter into you, and ye shall live:

[6] And I will lay sinews upon you, and will bring up flesh upon you, and cover you with skin, and put breath in you, and ye shall live; and ye shall know that I am the LORD.

[7] So I prophesied as I was commanded: and as I prophesied, there was a noise, and behold a shaking, and the bones came together, bone to his bone.

[8] And when I beheld, lo, the sinews and the flesh came up upon them, and the skin covered them above: but there was no breath in them.

[9] Then said he unto me, Prophesy unto the wind, prophesy, son of man, and say to the wind, Thus saith the Lord GOD; Come from the four winds, O breath, and breathe upon these slain, that they may live.

[10] So I prophesied as he commanded me, and the breath came into them, and they lived, and stood up upon their feet, an exceeding great army.

[11] Then he said unto me, Son of man, these bones are the whole house of Israel: behold, they say, Our bones are dried, and our hope is lost: we are cut off for our parts.

[12] Therefore prophesy and say unto them, Thus saith the Lord GOD; Behold, O my people, I will open your graves, and cause you to come up out of your graves, and bring you into the land of Israel.

¹³ And ye shall know that I am the LORD, when I have opened your graves, O my people, and brought you up out of your graves,

¹⁴ And shall put my spirit in you, and ye shall live, and I shall place you in your own land: then shall ye know that I the LORD have spoken it, and performed it, saith the LORD.

¹⁵ The word of the LORD came again unto me, saying,

¹⁶ Moreover, thou son of man, take thee one stick, and write upon it, For Judah, and for the children of Israel his companions: then take another stick, and write upon it, For Joseph, the stick of Ephraim and for all the house of Israel his companions:

¹⁷ And join them one to another into one stick; and they shall become one in thine hand.

¹⁸ And when the children of thy people shall speak unto thee, saying, Wilt thou not shew us what thou meanest by these?

¹⁹ Say unto them, Thus saith the Lord GOD; Behold, I will take the stick of Joseph, which is in the hand of Ephraim, and the tribes of Israel his fellows, and will put them with him, even with the stick of Judah, and make them one stick, and they shall be one in mine hand.

²⁰ And the sticks whereon thou writest shall be in thine hand before their eyes.

²¹ And say unto them, Thus saith the Lord GOD; Behold, I will take the children of Israel from among the heathen, whither they be gone, and will gather them on every side, and bring them into their own land:

²² And I will make them one nation in the land upon the mountains of Israel; and one king shall be king to them all: and they shall be no more two nations, neither shall they be divided into two kingdoms any more at all.

²³ Neither shall they defile themselves any more with their idols, nor with their detestable things, nor with any of their transgressions: but I will save them out of all their dwelling-places, wherein they have sinned, and will cleanse them: so shall they be my people, and I will be their God.

²⁴ And David my servant shall be king over them; and they all shall have one shepherd: they shall also walk in my judgments, and observe my statutes, and do them.

[25] And they shall dwell in the land that I have given unto Jacob my servant, wherein your fathers have dwelt; and they shall dwell therein, even they, and their children, and their children's children for ever: and my servant David shall be their prince for ever.

[26] Moreover I will make a covenant of peace with them; it shall be an everlasting covenant with them: and I will place them, and multiply them, and will set my sanctuary in the midst of them for evermore.

[27] My tabernacle also shall be with them: yea, I will be their God, and they shall be my people.

[28] And the heathen shall know that I the LORD do sanctify Israel, when my sanctuary shall be in the midst of them for evermore. Ezekiel 37

Scholars agree that the dry bones are Israel and this chapter depicts them becoming a nation again. It is a 70-year term in Jerimiah along with the prophecy in Leviticus that takes us to the year 1948.[222] The math becomes complicated, but it is enough to suffice that it is to the year. But, the explanation is that because of the rebellion of Israel, the time was extended. So now when we read about Israel becoming a nation, and the time of the return of the Messiah, it is imminent at any time now! Consider this Scripture:

[32] Now learn a parable of the fig tree; When his branch is yet tender, and putteth forth leaves, ye know that summer is nigh:

[33] So likewise ye, when ye shall see all these things, know that it is near, even at the doors.

[34] Verily I say unto you, This generation shall not pass, till all these things be fulfilled. Matthew 24: 32 - 34

---

[222] Ibid.

So, in this Scripture, it is considered that the fig tree is the Nation of Israel. And if the generation will not pass (away) before everything is fulfilled, that would generally mean that not everyone alive at the time when Israel became a nation again will pass away. Given that the maximum age is 120 years for normal people, which should mean that by the year of 2068, the Rapture should occur! But I am by no means setting any specific date, and this calculation could indeed be totally wrong. It was generally accepted a generation might be 70 years or less, but that time has now passed, so if we take it to the extreme, we will have 47 years from 2021 before the return of Jesus. It could happen at any time now, however! This whole idea is speculation, but it tells us the time is near. If it does not happen by 2068, I have a mistake in my math or misunderstanding of what the Scripture means about the fig tree not a problem with the Bible and Jesus. It is logical to assume, though, that what Jesus said is what it means. Be prepared for any day for the return of Jesus! There is nothing, according to biblical scholars that must happen now before the Rapture. Some people may contend the Temple needs to be rebuilt, but that can occur *after* the Rapture, and in my mind, it will. Right now we have the Islamic people controlling the Temple Mount, and they are not going to lay down and let Israel build a new Temple on the place they have their mosque. If, however, it was aliens that demanded it be destroyed and a new Temple built, they most likely would be powerless to resist, or if they did, they would be eliminated by these Demonic Alien Overlords. It would have to be something this earth-shattering to produce such a change. The disclosure of so-called alien creatures would be the catalyst that propels the Temple to be rebuilt! With the alien technological superiority, the enemies of Israel would not resist such an order, I surmise. Nothing else I can see will cause such a radical change in the minds and hearts of the enemies of Israel.

How do we come up with the scenario of the Rapture occurring without the Temple being rebuilt? It is gleaned from Bible study by many scholars much smarter than me. Consider this outline of what is proposed to happen in biblical eschatology:

1. Rapture
2. First half of the Tribulation
3. Second half of the Tribulation
4. Battle of Armageddon
5. Marriage of the Lamb (Second Coming)
6. Glorious Return of Jesus with the Church and His Angels and Judgment of Believers
7. Antichrist and the False Prophet cast into the Lake of Fire
8. Satan Bound at the beginning of the New Millennium
9. The New Millennium
10. Satan loosed at the end of the New Millennium for a short time and bound for eternity
11. Great White Throne Judgement of the Wicked
12. Antichrist and False Prophet cast into the Lake of Fire
13. New Heaven
14. New Earth
15. New Jerusalem
16. New Temple
17. New Light

18. New Paradise[223]

Nothing else must happen now before the Rapture! All prophecy concerning the Return of Jesus is fulfilled, then everything else takes place after that. The Temple, I contend, is rebuilt in the first half of the Great Tribulation, although I have no proof or incite into this, beyond what it is like today. The Demonic Aliens must be disclosed, or something just as earth-shattering, before it would happen. So, the timeline leading up to the Rapture and the rebuilding of the Temple is this:

Rapture of the Church

Disclosure of the Demonic Alien Overlords with Satan himself as their ruler.

The mortal wound of the Antichrist is healed after his head is cut off and because of this, he is considered to be the Messiah.

A peace treaty is signed between Israel and the Demonic Alien Overlords which allows the Temple to be rebuilt in the first half of the Great Tribulation.

After the Temple is rebuilt, the Antichrist walks into the Holy of Holies and declares himself to be God, which of course he is not.

Then begins the most terrible time on the Earth with the second half of the Great Tribulation.

During all this time, it is required for everyone to get the implant (Mark of the Beast), or they are rounded up and put in FEMA camps and killed by guillotine if they refuse the implant.

All these things that are expected to happen from 1 – 7 above are utter speculation on my part, but it makes sense that it could and will happen just like this. I have

---

[223] Gleaned from several charts in the Tim LaHaye Prophecy Study Bible published by AMG Publishers © 2000.

no proof or biblical knowledge beyond what everyone else has but am impressed that it will happen just like this. If I am totally wrong and all this speculation is just that, it does not change what is happening today and the relevancy of the Bible and the Return of Jesus. Consider the disclosure of UFOs being real by our government now. This means the time is even shorter than we think for them to be fully disclosed. When that happens, the world will worship them and not God! How did I come up with this idea that Demonic Alien Overlords are going to try to rule over us? They have the technology to do so, they have the craft that can travel in water, air, and space at speeds beyond our capabilities. They will implement the implants that everyone will be required to take, which is what it says in Revelation. We are powerless to stop them when they want to take over the world. They must be demonic, and we must be prepared. People will throw away their Bibles because they will believe the Bible is wrong when the aliens arrive. I say aliens, but **again they are not aliens but demons.**

But, since the Bible doesn't mention them, why would we think they are even involved with the world and the end times? **You are missing the point!** Again, I must keep reminding myself that they are demons and not aliens. When you substitute the word demons for aliens in this case, these creatures fit perfectly in all aspects of historical times when cultures were influenced by such creatures, and even today when we are being influenced to trash God and go with them. Use the term DAO (Demonic Alien Overlords) now whenever you hear UFO or UAP (Unidentified Aerial Phenomenon) and you will begin to change your mindset. Whatever term they apply to them, just say demons, and if you do that you will know what they are!

I have placed myself in the "Crazy" file with trying to describe how they get us to sin. The concept they are shooting down "waves of sin" from above us is just strange. I am trying to answer, for myself, why I can't stop sinning. No one can, and we are distressed to our limits. Stop doing any sin at all and prove me wrong! You lie to your wife or husband all the time. You yell at your kids for no reason. You have bad sexual thoughts or angry thoughts. You, even as a Christian can't stop all sin in your life, so don't judge me either. Explain for yourself what these Demonic Alien Overlords are doing to you to influence you to sin, and you have the answer. If you think becoming a Christian means you won't ever sin again, you are wrong. Even Paul the strongest Christian yet tells us:

¹³ **Was then that which is good made death unto me? God forbid. But sin, that it might appear sin, working death in me by that which is good; that sin by the commandment might become exceeding sinful.**
¹⁴ **For we know that the law is spiritual: but I am carnal, sold under sin.**
¹⁵ **For that which I do I allow not: for what I would, that do I not; but what I hate, that do I.**
¹⁶ **If then I do that which I would not, I consent unto the law that it is good.**
¹⁷ **Now then it is no more I that do it, but sin that dwelleth in me.**
¹⁸ **For I know that in me (that is, in my flesh,) dwelleth no good thing: for to will is present with me; but how to perform that which is good I find not.**
¹⁹ **For the good that I would I do not: but the evil which I would not, that I do.**
²⁰ **Now if I do that I would not, it is no more I that do it, but sin that dwelleth in me.**
²¹ **I find then a law, that, when I would do good, evil is present with me.** Romans 7: 13 -21

Even Paul, could not keep from sin in his life, so join the club of Christians now being distressed and influenced by these Demonic Alien Overlords! Realize it will get worse before it gets better. One thing I have found is that if you are of no threat to them as a mediocre Christian, they leave you alone, or if they can't make you sin, even as a non-Christian, the same is true.

Chapter 25
CHARIOTS OF GOD

I stated unequivocally that the Bible does not mention aliens, but it does mention UFOs, so people will assume that this means aliens from another planet, but I will show why this is not at all the case. UFOs are not aliens and these crafts have no bearing on what aliens are or are not. Given that angels exist, they can also travel in supernatural craft to our knowledge base and be just like the aliens we see in our skies, but not be these demonic creatures. The best-known example of a UFO sighting which is of God and an angelic nature comes of course from Ezekiel where he observes a "wheel within a wheel," but this is not a demonic craft, but an angelic craft from God. The Scripture I am referring to is here in chapter 10 of Ezekiel which describes him seeing a vision of a "chariot of God" which he says had a "wheel within a wheel." This vision, however, is not of these creatures which plague us and is clearly from God and not demons. Notice he sees cherubs (angels of God) that have different images of first a face of a cherub, then a face of a man, then the face of a lion, and lastly the face of an eagle. From this image, we see the "glory of the Lord" shown out of it. This is a difficult Scripture to analyze with our limited viewpoint of what he is describing, but just know it is from God and not from these demonic entities. How do we know it is from God and not the Demonic Aliens we are observing today? Let's analyze what we see here in this Scripture with what is happening today. From what we know about what we see today, how is this vision different?

True Aliens *are not* mentioned in the Bible, therefore do not likely exist.

No aliens are mentioned here, just a UFO-type craft, so this is not an alien craft.

We could not have evolved, therefore, neither could aliens.

Nothing notates any aliens being "evolved" so this is still not notating aliens.

Demons *are* mentioned in the Bible; therefore, any discovered so-called *aliens* are demons.

This craft is angelic and not demonic, so it is not referring to aliens.

The lie they will state is that they "seeded" our planet, thus taking credit for God's creation.

This angelic craft is from God and not demonic.

The bases they have are here on Earth and not on some other planet, but they will lie and say they are from another planet. Government officials have stated there is no indication UFOs or UAPs are from another planet.

This craft is from God and not from another planet

They are abducting people, doing medical experiments on us, collecting genetic material, and all manner of nefarious acts, even collecting sexual materials (sperm and eggs) therefore they are not benevolent aliens but are shown to be demons.

While Elijah was "taken up" by God it does not say by a UFO. Enock was also "taken up" but it does not say by any type of craft.

This collecting of genetic material would be for the purpose of making hybrid-alien creatures which is a continuation of what was done before in Genesis when the demons "mated" with human females and produced giants.

Ezekiel was not captured and released with genetic material being taken but he saw the craft that was like a chariot.

Angels are a little higher than us, thus fallen angels (demons) would be more technologically advanced than we are, as these creatures are.

The craft Ezekiel saw was of a nature not described by what we see today, therefore, it is not an alien craft.

When they are disclosed, people will worship them and forget about God, like the Bible says, the whole world will wonder after the beast.

Ezekiel does not state to worship the craft and their occupants; therefore, we are to worship only God and not the craft or the occupants of the craft.

They are implanting people with devices, which falls into place with the Bible where everyone will be required to get the mark to buy or sell, which the implant could in fact be a type of "mark."

No implant is discussed in this chapter about when Ezekiel saw the craft, therefore this is not of aliens but God.

Demonic entities control people and Demonic Alien Overlords flying above us would be in a position to do just that.

The creatures in the craft did not control him, but it was a vision of God and not of demons.

Other reasons demonstrating a young Universe.

No indication is given that these creatures in the craft are aliens, so it was of God and not of alien control.

They are sexually depraved.

The craft had no sexual context or depravity, but was displaying the Glory of God, therefore, is from God and angelic and not demonic.

They are not loving, therefore, are not of God!

This craft had an angelic loving nature, so this is of God and not aliens.

¹Then I looked, and, behold, in the firmament that was above the head of the cherubims there appeared over them as it were a sapphire stone, as the appearance of the likeness of a throne.

² And he spake unto the man clothed with linen, and said, Go in between the wheels, even under the cherub, and fill thine hand with coals of fire from between the cherubims, and scatter them over the city. And he went in in my sight.

³ Now the cherubims stood on the right side of the house, when the man went in; and the cloud filled the inner court.

⁴ Then the glory of the LORD went up from the cherub, and stood over the threshold of the house; and the house was filled with the cloud, and the court was full of the brightness of the LORD's glory.

⁵ And the sound of the cherubims' wings was heard even to the outer court, as the voice of the Almighty God when he speaketh.

⁶ And it came to pass, that when he had commanded the man clothed with linen, saying, Take fire from between the wheels, from between the cherubims; then he went in, and stood beside the wheels.

⁷ And one cherub stretched forth his hand from between the cherubims unto the fire that was between the cherubims, and took thereof, and put it into the hands of him that was clothed with linen: who took it, and went out.

⁸ And there appeared in the cherubims the form of a man's hand under their wings.

⁹ And when I looked, behold the four wheels by the cherubims, one wheel by one cherub, and another wheel by another cherub: and the appearance of the wheels was as the colour of a beryl stone.

¹⁰ And as for their appearances, they four had one likeness, as if a wheel had been in the midst of a wheel.

¹¹ When they went, they went upon their four sides; they turned not as they went, but to the place whither the head looked they followed it; they turned not as they went.

¹² And their whole body, and their backs, and their hands, and their wings, and the wheels, were full of eyes round about, even the wheels that they four had.

¹³ As for the wheels, it was cried unto them in my hearing, O wheel.

¹⁴ And every one had four faces: the first face was the face of a cherub, and the second face was the face of a man, and the third the face of a lion, and the fourth the face of an eagle.

¹⁵ And the cherubims were lifted up. This is the living creature that I saw by the river of Chebar.

¹⁶ And when the cherubims went, the wheels went by them: and when the cherubims lifted up their wings to mount up from the earth, the same wheels also turned not from beside them.

¹⁷ When they stood, these stood; and when they were lifted up, these lifted up themselves also: for the spirit of the living creature was in them.

¹⁸ Then the glory of the LORD departed from off the threshold of the house, and stood over the cherubims.

¹⁹ And the cherubims lifted up their wings, and mounted up from the earth in my sight: when they went out, the wheels also were beside them, and every one stood at the door of the east gate of the LORD's house; and the glory of the God of Israel was over them above.

²⁰ This is the living creature that I saw under the God of Israel by the river of Chebar; and I knew that they were the cherubims.

²¹ Every one had four faces apiece, and every one four wings; and the likeness of the hands of a man was under their wings.

²² And the likeness of their faces was the same faces which I saw by the river of Chebar, their appearances and themselves: they went every one straight forward.

We also have the Scripture that states that Enock was "taken" by God, but it does not say by what type of craft took him or even if there was a craft. It is stated here:

> **And Enoch walked with God: and he was not; for God took him.** Genesis 5:24

He was "taken" by God, but it does not mention *how* he was taken. Another person that was "taken" by God was Elijah. It describes the craft as a "chariot of fire" or a whirlwind. This is not a UFO.

Another instance when we see "chariots of fire" was when Elisha prayed and there were chariots of fire round about them:

> **And it came to pass, as they still went on, and talked, that, behold, there appeared a chariot of fire, and horses of fire, and parted them both asunder; and Elijah went up by a whirlwind into heaven.** 2 Kings 2:11

> [16] **And he answered, Fear not: for they that be with us are more than they that be with them.**
> [17] **And Elisha prayed, and said, LORD, I pray thee, open his eyes, that he may see. And the LORD opened the eyes of the young man; and he saw: and, behold, the mountain was full of horses and chariots of fire round about Elisha.**
> [18] **And when they came down to him, Elisha prayed unto the LORD, and said, Smite this people, I pray thee, with blindness. And he smote them with blindness according to the word of Elisha.** 2 Kings 6: 16 - 18

All the times when we see something like what we might think is talking about UFOs is not at all what we are seeing today. There are no indications that the UFOs people are seeing today are anything at all like these creatures, which are demons, and the "chariots of fire" in the Bible. The times when people were "taken up" was not by a UFO, that it mentions and so aliens are not mentioned in the Bible. There are other incidents when supernatural occurrences happened that might be construed to be chariots of God or the Lord standing in the way of things happening. One such incident is with Balaam and the donkey when the Angel of the Lord blocked his way. This again does not depict some type of UFO. The Scripture states:

²¹ And Balaam rose up in the morning, and saddled his ass, and went with the princes of Moab.

²² And God's anger was kindled because he went: and the angel of the LORD stood in the way for an adversary against him. Now he was riding upon his ass, and his two servants were with him.

²³ And the ass saw the angel of the LORD standing in the way, and his sword drawn in his hand: and the ass turned aside out of the way, and went into the field: and Balaam smote the ass, to turn her into the way.

²⁴ But the angel of the LORD stood in a path of the vineyards, a wall being on this side, and a wall on that side.

²⁵ And when the ass saw the angel of the LORD, she thrust herself unto the wall, and crushed Balaam's foot against the wall: and he smote her again.

²⁶ And the angel of the LORD went further, and stood in a narrow place, where was no way to turn either to the right hand or to the left.

²⁷ And when the ass saw the angel of the LORD, she fell down under Balaam: and Balaam's anger was kindled, and he smote the ass with a staff.

²⁸ And the LORD opened the mouth of the ass, and she said unto Balaam, What have I done unto thee, that thou hast smitten me these three times?

²⁹ And Balaam said unto the ass, Because thou hast mocked me: I would there were a sword in mine hand, for now would I kill thee.

This could obviously not be construed to be a UFO but is showing how things can be hidden from our sight, angelic or demonic. Demons always hide! Angels when they appear, they show themselves. From this chapter we can discern those aliens are not really mentioned in the Bible and UFOs, are also in dispute, as to whether what Ezekiel saw was in fact a UFO or the Glory of God. In any event, we do not have a clear representation of any indication there are aliens, but still, plenty of indications there are demons.

Chapter 26
ANGELS VS DEMONIC ALIEN OVERLORDS

I keep calling these creatures either Demonic Aliens or Demonic Alien Overlords. I hope you glean from this book what I mean by this. These demons would take over our world and control us like cattle. Although I don't have a reference, I have heard it was one-third of the angels that rebelled with Satan, and these became the fallen angels (Demonic Aliens) we are seeing today. They were banished out of Heaven, so they now dwell on the Earth. A Scripture supporting this is found here:

> **[3] And there appeared another wonder in Heaven; and behold a great red dragon, having seven heads and ten horns, and seven crowns upon his heads.**
> **[4] And his tail drew the third part of the stars of Heaven, and did cast them to the earth: and the dragon stood before the woman which was ready to be delivered, for to devour her child as soon as it was born.** Revelation 12:3 - 4

The third part of the stars could indicate a third of the angels, but it does not state that. But where did these Demonic Alien Overlords come from? The Bible tells us this:

> **[7] And there was war in Heaven: Michael and his angels fought against the dragon; and the dragon fought and his angels,**
> **[8] And prevailed not; neither was their place found any more in Heaven.**
> **[9] And the great dragon was cast out, that old serpent, called the Devil, and Satan, which deceiveth the whole world: he was cast out into the earth, and his angels were cast out with him.** Revelation 12: 7 - 9

How there could have been a war in Heaven, it does not say, but because there was, we are now faced with fighting the devil and his angels here, and these are the

Demonic Alien Overlords I am talking about. You might just choose to call them demons but be aware they could in fact travel in crafts like the UFOs and by now have humanoid bodies that they made. Because of their high technology, and lying and saying they are from outer space, the government is "in bed" with them and helping them to develop more humanoid bodies in exchange for their technology (although I have no proof). When the Roswell Crash happened and they discovered bodies, the rumor is at least one of them survived and with this we now have them collaborating with the government and the Deep State to dictate what we do, through mandates from the government. We are required to wear masks and it is required to be vaccinated to even work at many jobs now. Angels would not be mandating such actions to us, but Demonic Aliens certainly would. We can see how it all started at the very beginning when Satan lied to Eve and caused her to sin, that he is a liar and so are all of the other demons that are troubling us here on this Earth. The familiar Scripture reads:

> **3 Now the serpent was more subtil than any beast of the field which the LORD God had made. And he said unto the woman, Yea, hath God said, Ye shall not eat of every tree of the garden?**
> **2 And the woman said unto the serpent, We may eat of the fruit of the trees of the garden:**
> **3 But of the fruit of the tree which is in the midst of the garden, God hath said, Ye shall not eat of it, neither shall ye touch it, lest ye die.**
> **4 And the serpent said unto the woman, Ye shall not surely die:**
> **5 For God doth know that in the day ye eat thereof, then your eyes shall be opened, and ye shall be as gods, knowing good and evil.**

⁶ And when the woman saw that the tree was good for food, and that it was pleasant to the eyes, and a tree to be desired to make one wise, she took of the fruit thereof, and did eat, and gave also unto her husband with her; and he did eat.
⁷ And the eyes of them both were opened, and they knew that they were naked; and they sewed fig leaves together, and made themselves aprons.
⁸ And they heard the voice of the LORD God walking in the garden in the cool of the day: and Adam and his wife hid themselves from the presence of the LORD God amongst the trees of the garden. Genesis 3: 1 - 8

The lie in verse four "Ye shall not surely die," is going to be the same lie when the evil Demonic Alien Overlords give us the implant to take. We will go to Hell if we take it, but they will leave that part out, of course! It is all through the Bible how demons try to control people and other angels, most notably when Satan tempted Christ and tried to get him to give up his claim to the world. The Scripture states it like this:

¹ Then was Jesus led up of the Spirit into the wilderness to be tempted of the devil.
² And when he had fasted forty days and forty nights, he was afterward an hungred.
³ And when the tempter came to him, he said, If thou be the Son of God, command that these stones be made bread.
⁴ But he answered and said, It is written, Man shall not live by bread alone, but by every word that proceedeth out of the mouth of God.
⁵ Then the devil taketh him up into the holy city, and setteth him on a pinnacle of the temple,
⁶ And saith unto him, If thou be the Son of God, cast thyself down: for it is written, He shall give his angels charge concerning thee: and in their hands they shall bear thee up, lest at any time thou dash thy foot against a stone.

> ⁷ Jesus said unto him, It is written again, Thou shalt not tempt the Lord thy God.
> ⁸ Again, the devil taketh him up into an exceeding high mountain, and sheweth him all the kingdoms of the world, and the glory of them;
> ⁹ And saith unto him, All these things will I give thee, if thou wilt fall down and worship me.
> ¹⁰ Then saith Jesus unto him, Get thee hence, Satan: for it is written, Thou shalt worship the Lord thy God, and him only shalt thou serve.
> ¹¹ Then the devil leaveth him, and, behold, angels came and ministered unto him. Matthew 4: 1 - 11

We can see from this Scripture that Satan uses every trick in his playbook to try to get us to do what he wants, but with Jesus, he failed. We have to follow this example to combat the devil and the demons that will reveal themselves as so-called aliens if we want to survive to the end. If they are revealed before the Rapture, be prepared to fight! Find the answers on how to fight them with Scripture in the Bible. Whatever they say, combat it with the truth from the Bible!

Whenever true angels appear, however, we will know what they are doing is in fact of God! They will appear to us, even as a normal-looking man, and tell us good things, which we will know are true because we can test their words from the Bible. An example of this is of course when the Angel Gabriel announced to Mary what would happen. That was already prophesied in the Bible and foretold in Isaiah here:

> **Therefore, the Lord himself shall give you a sign; Behold, a virgin shall conceive, and bear a son, and shall call his name Immanuel.** Isaiah 7:14

So when the Angel Gabriel came to Mary and declared what would happen, we could check it against what is in the Bible and know it is true. As it states here:

²⁶ And in the sixth month the angel Gabriel was sent from God unto a city of Galilee, named Nazareth,
²⁷ To a virgin espoused to a man whose name was Joseph, of the house of David; and the virgin's name was Mary.
²⁸ And the angel came in unto her, and said, Hail, thou that art highly favoured, the Lord is with thee: blessed art thou among women.
²⁹ And when she saw him, she was troubled at his saying, and cast in her mind what manner of salutation this should be.
³⁰ And the angel said unto her, Fear not, Mary: for thou hast found favour with God.
³¹ And, behold, thou shalt conceive in thy womb, and bring forth a son, and shalt call his name JESUS.
³² He shall be great, and shall be called the Son of the Highest: and the Lord God shall give unto him the throne of his father David:
³³ And he shall reign over the house of Jacob for ever; and of his kingdom there shall be no end.
³⁴ Then said Mary unto the angel, How shall this be, seeing I know not a man?
³⁵ And the angel answered and said unto her, The Holy Ghost shall come upon thee, and the power of the Highest shall overshadow thee: therefore also that holy thing which shall be born of thee shall be called the Son of God.
³⁶ And, behold, thy cousin Elisabeth, she hath also conceived a son in her old age: and this is the sixth month with her, who was called barren.
³⁷ For with God nothing shall be impossible. Luke 1: 26 - 37

This was foretold in the Bible how a virgin will conceive. We also see where there was the time when the Angel Gabriel came to Zacharias while he was performing his duties in the Temple. It tells us:

> ¹³ **But the angel said unto him, Fear not, Zacharias: for thy prayer is heard; and thy wife Elisabeth shall bear thee a son, and thou shalt call his name John.**
> ¹⁴ **And thou shalt have joy and gladness; and many shall rejoice at his birth.**
> ¹⁵ **For he shall be great in the sight of the Lord, and shall drink neither wine nor strong drink; and he shall be filled with the Holy Ghost, even from his mother's womb.**
> ¹⁶ **And many of the children of Israel shall he turn to the Lord their God.**
> ¹⁷ **And he shall go before him in the spirit and power of Elias, to turn the hearts of the fathers to the children, and the disobedient to the wisdom of the just; to make ready a people prepared for the Lord.**
> ¹⁸ **And Zacharias said unto the angel, Whereby shall I know this? for I am an old man, and my wife well stricken in years.**
> ¹⁹ **And the angel answering said unto him, I am Gabriel, that stand in the presence of God; and am sent to speak unto thee, and to shew thee these glad tidings.**
> ²⁰ **And, behold, thou shalt be dumb, and not able to speak, until the day that these things shall be performed, because thou believest not my words, which shall be fulfilled in their season.** Luke 1: 13 - 20

We should also take heed to honor the Word of God when it is telling us something or be prepared to suffer the consequences with perhaps hardship until we realize what it is saying. In the case of Zacharias, he was made dumb, because he failed to heed the word from the Angel Gabriel. In our case, when the Bible talks about the end times, and with the current events that are taking place, there will be a great deception. I go back to the Scripture that tells us:

> [1] And I stood upon the sand of the sea, and saw a beast rise up out of the sea, having seven heads and ten horns, and upon his horns ten crowns, and upon his heads the name of blasphemy.
> [2] And the beast which I saw was like unto a leopard, and his feet were as the feet of a bear, and his mouth as the mouth of a lion: and the dragon gave him his power, and his seat, and great authority.
> [3] And I saw one of his heads as it were wounded to death; and his deadly wound was healed: and all the world wondered after the beast.
> [4] And they worshipped the dragon which gave power unto the beast: and they worshipped the beast, saying, Who is like unto the beast? who is able to make war with him?
> [5] And there was given unto him a mouth speaking great things and blasphemies; and power was given unto him to continue forty and two months.
> [6] And he opened his mouth in blasphemy against God, to blaspheme his name, and his tabernacle, and them that dwell in heaven.
> [7] And it was given unto him to make war with the saints, and to overcome them: and power was given him over all kindreds, and tongues, and nations.
> [8] And all that dwell upon the earth shall worship him, whose names are not written in the book of life of the Lamb slain from the foundation of the world. Revelation 13: 1 - 8

It tells us in verse 7 that he makes war with the Saints. That means at least some Christians will be here when all this happens! We would hope and pray that it was not so, that the Rapture removes all the Christians, but it appears there will be some still here, either that became Christians after the Rapture, or that the Rapture occurs Mid-Tribulation. Either way, the great deception would be these Demonic Aliens masquerading as aliens and not as demons. There is a myriad of examples where angels from God appear in our midst, even today, and although we may not even recognize them, they are real and always do good from God. One other example in Scripture was when the donkey saw the Angel of the Lord and Balaam did not and he suffered consequences from the Angel of the Lord as it depicts here:

> *[22] And God's anger was kindled because he went: and the angel of the LORD stood in the way for an adversary against him. Now he was riding upon his ass, and his two servants were with him.*
> *[23] And the ass saw the angel of the LORD standing in the way, and his sword drawn in his hand: and the ass turned aside out of the way, and went into the field: and Balaam smote the ass, to turn her into the way.*
> *[24] But the angel of the LORD stood in a path of the vineyards, a wall being on this side, and a wall on that side.*
> *[25] And when the ass saw the angel of the LORD, she thrust herself unto the wall, and crushed Balaam's foot against the wall: and he smote her again.*
> *[26] And the angel of the LORD went further, and stood in a narrow place, where was no way to turn either to the right hand or to the left.*
> *[27] And when the ass saw the angel of the LORD, she fell down under Balaam: and Balaam's anger was kindled, and he smote the ass with a staff.*
> *[28] And the LORD opened the mouth of the ass, and she said unto Balaam, What have I done unto thee, that thou hast smitten me these three times?*
> *[29] And Balaam said unto the ass, Because thou hast mocked me: I would there were a sword in mine hand, for now would I kill thee.*
> *[30] And the ass said unto Balaam, Am not I thine ass, upon which thou hast ridden ever since I was thine unto this day? was I ever wont to do so unto thee? and he said, Nay.*
> *[31] Then the LORD opened the eyes of Balaam, and he saw the angel of the LORD standing in the way, and his sword drawn in his hand: and he bowed down his head, and fell flat on his face.*
> *[32] And the angel of the LORD said unto him, Wherefore hast thou smitten thine ass these three times? behold, I went out to withstand thee, because thy way is perverse before me:*

<blockquote>
³³ *And the ass saw me, and turned from me these three times: unless she had turned from me, surely now also I had slain thee, and saved her alive.*
³⁴ *And Balaam said unto the angel of the LORD, I have sinned; for I knew not that thou stoodest in the way against me: now therefore, if it displease thee, I will get me back again.* Numbers 22: 22 - 35
</blockquote>

Angels always do good, but we may not even be aware of what they are doing. Demonic Alien Overlords, although they may seem good, are liars, evil demons that will attempt to get you to do what you know is wrong. When they are disclosed, know they are evil and not good. Right now, they are obviously limited by God in what they are able to do, but as it says in Revelation, soon the restrictions will be taken away, and we will be at their mercy!

Chapter 27
HOW THE (ALIEN) IMPLANT WILL CHANGE US

Even today you can get an implant in your hand or another place on your body. The authorities will say it has nothing to do with what it talks about in Revelation, but what is it for? They put all your financial information on it so you can do financial transactions without an external card or check. Some employers have gone so far as to require their employees to have the implant to work for their company, so they can enter the building and secure areas.

On 1 August 2017, workers at Three Square Market, a Wisconsin-based company specializing in vending machines, lined up in the office cafeteria to be implanted with microchips. One after the other, they held out a hand to a local tattoo artist who pushed a rice-grain sized implant into the flesh between the thumb and forefinger. The 41 employees who opted into the procedure received complimentary t-shirts that read "I Got Chipped".[224]

People today will think it is "cool" to have it because you don't need to carry any ID. The current type of implant is about the size of a grain of rice.[225] In the beginning, it will appear like everything is fine, then it begins to change us. But this creates limitless worker surveillance.[226] Not only this but this is indeed a change in their

---

[224] "The rise of microchipping: are we ready for technology to get under the skin?"
https://www.theguardian.com/technology/2019/nov/08/the-rise-of-microchipping-are-we-ready-for-technology-to-get-under-the-skin
Accessed 8.20.2021

[225] Ibid.

[226] Ibid.

body that will result in them being monitored from external sources for other things, like financial, health, etc. Later they won't need a credit card or debit card to shop. They will just wave their hand over the terminal at the checkout and the funds will be deducted from their account(s). Everything will seem fine for a little while, but then things will begin to change, even with this type of implant. While it is not alien and made by human hands the effects will become known later about how it will affect us. Consider this passage from Revelation chapter 10:

[10] And the fifth angel poured out his vial upon the seat of the beast; and his kingdom was full of darkness; and they gnawed their tongues for pain,
[11] And blasphemed the God of heaven because of their pains and their sores, and repented not of their deeds. Revelation 16: 10 - 11

The culprit that causes such a thing to happen is indeed the implant, either the implant men have made or the implant the Demonic Aliens will require us to get when they are revealed. Right now, the people who took this implant have very little discomfort. They are assured everything will be fine. They keep their jobs, go home to their families and everything seems "hunky-dory." Then later they will begin to have complications. Whenever we can implant something under someone's skin, we can do things to them and with them, they are not aware of. I heard the objective was to have everyone in the world micro-chipped by a certain date and then if they do anything wrong, you just shut their chip off! They can't buy food anymore, and therefore can't eat. Where have we heard this before? It sounds like Revelation come to life!

<blockquote>
¹⁶ **And he causeth all, both small and great, rich and poor, free and bond, to receive a mark in their right hand, or in their foreheads:**

¹⁷ **And that no man might buy or sell, save he that had the mark, or the name of the beast, or the number of his name.**

¹⁸ **Here is wisdom. Let him that hath understanding count the number of the beast: for it is the number of a man; and his number is Six hundred threescore and six.** Revelation 13: 16 - 18
</blockquote>

This is absolutely the same scenario we see happening right before our eyes here and now, but when the Demonic Aliens introduce their implant, it will be much worse! First, it will heal all the diseases people have. They will tell us we won't die (a lie from Hell) because we will all be condemned to Hell if we take the implant. Then later it will begin where they will have the sores all over their bodies and become like zombies following the demons and doing everything they want and be powerless to resist.

<blockquote>
¹⁵ **And the kings of the earth, and the great men, and the rich men, and the chief captains, and the mighty men, and every bondman, and every free man, hid themselves in the dens and in the rocks of the mountains;**

¹⁶ **And said to the mountains and rocks, Fall on us, and hide us from the face of him that sitteth on the throne, and from the wrath of the Lamb:**

¹⁷ **For the great day of his wrath is come; and who shall be able to stand?** Revelation 6: 12 - 17
</blockquote>

It will be true that people who consent to the implant can't die, but the cost will be the sores and "gnawing or tongues for pain." (Revelation 16:10). In the end, all who take the implant, either from man or the demons (I contend) will be condemned to Hell!

Some people may say this implant, made by man, is not the Mark of the Beast. That may be true, but I don't want to take that chance, do you? It may just be the precursor to the real deal that will be implemented by the demons when they are fully disclosed and say they are aliens. People will rush to get the implant that cures their diseases, and then it will turn on them. People will be controlled and ruled by demons after they take the implant they provide.

First, it may be voluntary if you want it or not, then just like the Covid-19 vaccine, if you want to work at certain jobs, you must take it. Even today in New York City you are unable to sit down and eat at restaurants without a vaccination card. They are even talking also about not letting you go into a store to buy food without one! Then in the end everyone will be required to take it, or you will have no way to buy food and you will starve. Sounds like the "good" demonic plan is in place, the trigger is the Rapture, or even if they just arrive at the White House grounds!

Is there some type of "marker" in the vaccine? There was talk about putting one in there. I heard Bill Gates did it. I don't believe there is one in there, but still, I don't want to take that chance. What it is though, is a way to tell who is going to be "trouble" for the demons and the government will be when they implement mandatory chip implants. All the governments of the world will be involved to chip everyone, and there will be a One World Government that will implement the chip implants. There is already in place, the FEMA camps in the United States to round up people who won't take the implant, and guillotines in place to execute those who won't do it! I wish it were not happening, but I see it getting closer by the day!

The whole idea is control. You control people by not letting them eat. Thus, when you can't buy or sell, you starve, unless you can grow your own food, but they will also stop that. You will have no choice but to starve. So, for the most part, once you see your kids and your wife starving, you will go along with the implant to get them food, and everyone that takes it will be condemned to Hell, I am sorry to say. Then in the future, it begins to change you into a "cyborg" being, no longer human anymore, and all the sores and pain breaks out in your body. The experiments with the implant have already begun, so we are on the verge of what it talks about in Revelations!

But how does this little implant really change us? It's a mental thing. What God did is not good enough, we need more. We can become like a computer by implanting chips in our brain that allow us to process math like a computer. We can hook ourselves up to the Internet some way and know all knowledge, rather than to have to ask the computer. But, once we do that, while we know everything about everything, so does the world know everything about everything about us! We are now "caught" by these Demonic Alien Overlords and there is no escape from their grasp! Once we choose to go into this "Brave New World" we are condemned and there is no way to be delivered anymore!

In our world today, even I use my computer to type this book. I ask it questions to find answers to put down, I talk to my cell phone, and it tells me the answer. Today, for sure people are working to connect themselves in some way to the Internet directly and become a total cyborg with no more human characteristics at all! That is what the demonic implant will do, and people will go along blindly because they will think the demons are aliens and not demons and they are helping us to evolve. They just will not be aware of what they really are and will be fooled into submitting to their control, by their advanced technology, by the government's prodding them to comply and make everything better than what God did. They will totally deny God any credit, and place their trust in evolution and not God. That is the lie they have been drumming into children every day in school now. You were not created, you evolved, and the next step in human evolution is to become super-human like a computer!

A computer has no feeling. It doesn't worship God or anything. It just spits out data that was programmed in it. It does what the operator commands it too, thus with the implant, if you take it, you become unfeeling, unemotional, and a dead cyborg void of any emotions at all. Like the *Star Trek* television show, the cyborg were machines. We are not machines but are capable of emotion. Taking the implant would rid of our ability to have emotions, thus we would no longer be human. We would not worship God but be required to "worship" the devil by doing everything he requires us to do. Thus when it comes around to this time, there will be no problem putting people who refuse to take the implant under the guillotine and chop off their heads. People who do it will have no emotions, good or bad, about what they are doing, they are just "following orders!"

Think it won't happen this way? We already see a precursor with the NAZI party and the extermination attempt of the Jews. Even then, they were just following orders without emotion to gas the Jews and then burn them up. With the implant, people will be void of any emotions and will do whatever their orders are without any question. That is what I mean, we will no longer be human anymore, dead to God, and therefore condemned because we chose to become something we should not be, emotionless cyborgs that can't worship him, but only worship Satan! God is being pushed out every day by claiming he did not create us. They will not teach that we are created because they don't want it to be true. If there is a God, there is sin. I can't rob a business by shoplifting or fornicate without consequences, or get an abortion, or kill my wife, or any bad things, but must repent! I want to be God! Then I can do what I want, and do not have to repent or worship God, I can worship myself. But in the end, the choices I make will determine Heaven or Hell, either I believe in God or if I chose to not believe then I will go to Hell. The choice is also yours and yours alone, so please don't get the implant!

Chapter 28
CROP CIRCLES AND OTHER STRANGE THINGS

It was believed that the intricate crop circles being found in fields all over the world were attributed to aliens. They had very bizarre and seemingly impossible designs for people to be responsible for. This, however, for the most part, is false. Demons are not doing them, but men. This is just an example of people being led to do evil things that seem paranormal by demons, for no other reason than to do mischief. They have been appearing throughout history, primarily in England, and usually appear overnight.[227]. They were exposed as being done by people when it was found to be done by two brothers.

In 1991, two hoaxers, Doug Bower and Dave Chorley, took credit for having created many circles throughout England after one of their circles was described by an investigator as impossible for human beings to make.[228]

Hoaxes like these lend credence to people being impressed to do things for evil reasons and not for good. It is the same old story, we are fallen beings and, therefore, will do evil and mischievous deeds for no other reason than to be bad. It is not just the fact, that these men were destroying a farmer's crop by making circles using boards in the field, but also the fact that they were deceiving people. For the most part, then, crop circles are *not* being done by aliens. It is, however, the result of a demonic influence over people to deceive and lie about what they are doing.

---

[227] Wikipedia, "Crop Circle," https://en.wikipedia.org/wiki/Crop_circle Accessed 21.8. 2021
[228] Ibid.

There have been films showing some orbs of light appearing and then crop circles appearing below them, but these films have been discredited as fake. They are film trickery designed to place fear in people over something man is doing, I surmise. Even the most intricate designs require only advanced math to complete them, and it is obvious that anyone can draw out a design and transfer it into a field using simple ropes and a board. It may be that some are done by the Demonic Aliens, but I don't think that is the case. One commentator suggested there may be "clubs" that have contests to see who can make the most intricate designs. I think this is plausible and not the result of direct involvement by aliens, but still, a demonic influence that is the cause.

Even if crop circles are the result of Demonic Alien Overlords doing them, that still is not alien, but demonic. Everything that happens in your life and mine is either the result of God or the devil and here on the Earth, a lot of things that happen are the result of the devil. Take for example that you park your car in a parking lot and go to work and when you come back out the windows are smashed in, and people rummaged through your car to steel any loose change or other things you have in it. It wasn't aliens that did it obviously, but a person who was impressed by the devil to do evil. Then take the example of a person who finds a wallet with $100.00 in it, and

---

[229] Owned image from Shutterstock

returns it. That person is being impressed by God. There are two poles of actions, good and evil. God is good and the devil is evil. It will always be that way.

I thought I had witnessed a miracle one time, but it turned out not to be that. My wife's car had a busted taillight and one day I noticed it was fixed. I never did it! I even told my father that a miracle had happened. Then I recalled I took her car in to have the oil changed at a dealership, and it turned out the technician did it without charge! So much for the letdown. That dealership is not even in business anymore, I don't think, they did well so much they went out of business. But they were being led by God and not the devil. Things can get pretty dicey when we think about all the evil that is happening now. Our inner cities are now just zones no one goes to anymore to buy things. People run into the stores, fill up bags and walk right out without paying! They are not even prosecuting people if what they steel is under $1,000.00! How stupid! The business closes because their losses are going to be above their profits, and they can't stop the people from stealing. Now that is the devil working to destroy our society.

Other paranormal things we think we see are often illusions. Like the time I saw a very strange creature in the barn one morning. It had a head like an opossum and feet like a chicken. I ran and told dad and he brought a gun out and killed the opossum. There was no chicken involved. I saw something that wasn't really there. We may see things when it gets dark that are strange, but really they are normal.

BIGFOOT

One thing people have been seeing that I believe is real, however, is what is commonly known as Bigfoot or Sasquatch. It, however, is nothing more than an ape that lives in the United States and other areas. It really has no connection to the paranormal, although because it seems to just disappear, people want to assign UFOs to it as though that has anything at all to do with anything. There is an abundance of evidence to show it exists, like casts of footprints and even hair and blood samples that prove it exists but finding one alive or dead is a huge problem. They are smart and elusive. What they are is the remnants of a creature that lived earlier, called *Gigantopithecus* that lived at the beginning and is still alive today. This is another nail in the coffin for evolution because the creature they think died out a million or more years ago is still alive today with very little change! It is not supernatural or paranormal in any way, just another type of creature that God created. It is smart enough to stay away from humans, though. One day a dead one will be found and that will seal the deal about it being real, but for now, it is just an ape living in the United States and other wilderness areas of the world.

230

---

[230] Owned image from Shutterstock

THE LOCK NESS MONSTER

I'm sure you have heard of this one. It is, however, not paranormal or associated with UFOs, but is a remaining dinosaur, most likely known as a plesiosaur, which should also have died out about 66 million years ago,[231] according to evolution, but they are still alive today, which is another good nail in the coffin of evolutionary theory. There have been lots of people that have seen it, and scientific expeditions to prove it is real have shown pictures that very much resemble the ancient plesiosaur, still alive and thriving in Loch Ness in Scotland! It likely has a way to get out of the Lock through some type of cave system, into the ocean or other areas, or a place to winter, inside a cavern, so that is likely why it is still a mystery There are other places it can also be found living as well, such as Lake Champlain in New York. Here they call the creature Champ. While this may be another type of creature, it may in fact be the same.

[232]

MOKELE-MBEMBE

This is the name for a living dinosaur out of Africa which the local tribes have seen and even

---

[231] Wikipedia, "Plesiosauria," https://en.wikipedia.org/wiki/Plesiosauria Accessed 8.21.2.21

[232] Owned image from Shutterstock

have a name for.[233] It is like a Brachiosaurus, however not nearly as big. The creature lives in the swamps of Africa. Per Wikipedia:

Tales of entities like mokele-mbembe, living saurians walking around the African rain forest, are not rare, there have been multiple tales of large, smooth skinned quadrupeds with long necks that fed on large prey still living in central Africa. It was only after the description of the mokele-mbembe surfaced that the rest of the world started interpreting those legends as possessing a dinosaur-like body structure. A notable example would be the emela-ntouka, an elephant sized creature that shares a lot of similarities with the mokele-mbembe, it is described as having smooth skin, a strong and muscular tail and a "horn" or "tooth". Another similar creature, the jago-nini was described by Alfred Aloysius Smith who had worked for a British trading company in what is now Gabon in the late 1800s, briefly mentions it in his 1927 memoir. [234]

Other accounts have surfaced where creatures like this live in South America and other unexplored areas, where eyewitness accounts are numerous. These are obviously living dinosaurs, although not nearly as large anymore, that somehow survived the flood, or where their "kind" were actually taken on board the Ark. It is likely this type of creature would have been able to survive by being able to hold its breath for an extended period, like crocodiles, which would not, in my estimation have been taken on the Ark. Fish obviously were not taken on the Ark, and these creatures would possibly survive such a catastrophe as a World-wide Global Flood.

---

[233] Wikipedia, "Mokel-mbembe," https://en.wikipedia.org/wiki/Mokele-mbembe Accessed 8.21.2021
[234] Ibid.

235

All over the world, there are reports where people see what could be living dinosaurs, but they are brushed off as being false. One area where people are seeing the flying dinosaurs is Papua New Guinea which is very much unexplored. There have even been reports of people being killed by these creatures.[236] The local population calls it "The Demon Flyer" because it is so scary, and they are obviously aware of other creatures like bats that live on the large island, but this one is like a living dinosaur.[237]They have also been reported to have bioluminescence like lightning bugs and are primarily nocturnal.[238] I put this in here because if these creatures exist today, it knocks another hole in the theory of evolution and how these creatures could still exist today. Just like so many other living dinosaurs that have not evolved, here is another example of one still living today. The list of creatures that show no evolutionary change is very extensive and when you think about it, the obvious conclusion is

---

[235] Owned image from Shutterstock

[236] "MonsterQuest: Prehistoric Flying Monster," https://www.youtube.com/watch?v=dzb6ZWPq6AI&t=22s&ab_channel=HISTORY Accessed 9.7.2021

[237] Ibid

[238] Ibid.

there is no evolution, from one creature into another, of any species. I also covered this in my previous book *Finding Proof of Jesus*. There are literally hundreds!

Aardvark *(Orycteropus afer)*
Amami rabbit *(Pentalagus furnessi)*
Chevrotain*(Tragulidae)*
Elephant shrew*(Macroscelidea)*
Laotian rock rat *(Laonastes aenigmamus)*
Monito del Monte *(Dromiciops gliroides)*
Monotremes *(the* platypus *and* echidna*)*
Mountain beaver *(Aplodontia rufa)*
Okapi *(Okapia johnstoni)*
Opossums *(Didelphidae)*
Capybara *(Hydrochoerus hydrochaeris)*
Pygmy right whale *(Caperea marginata)*
Red panda *(Ailurus fulgens)*
Solenodon (Solenodon cubanus *and* Solenodonpara-doxus*)*
Shrew opossum *(Caenolestidae)*
Cetaceans
False killer whale *(Pseudorca crassidens)*
Birds
Pelicans *have been virtually unchanged sincethe* Eocene *and are noted to have been even more conservative across the Cenozoic than crocodiles.*
Acanthisittidae *(New Zealand "wrens")—two living species, a few more recently* extinct. *Distinct lineage of* Passeriformes.
Broad-billed sapayoa *(Sapayoa aenigma)— one living species. Distinct lineage of* Tyranni.
Bearded reedling *(Panurus biarmicus)—one living species. Distinct lineageof* Passerida *or* Sylvioidea.

Coliiformes *(mousebirds)—six living species in two genera. Distinct lineage of* Neoaves.
Hoatzin *(Ophisthocomus hoazin)—one living species. Distinct lineage of* Neoaves.
Magpie goose *(Anseranas semipalmata)—one living species. Distinct lineage of* Anseriformes.
Seriema *(Cariamidae)—two living species. Distinct lineage* Cariamae.
Tinamiformes *(tinamous) fifty living species Distinct Lineage* Palaeognathae
Reptiles
Alligator snapping turtle *(Macrochelys temminckii)*
Crocodilia *(crocodiles, gavials, caimans and* alligators*)*
Pig-nosed turtle *(Carettochelys insculpta)*
Snapping turtle *(Chelydra serpentina)*
Tuatara *(Sphenodon punctatus and* Sphenodon guntheri*)*
Amphibians
Giant salamanders *(*Cryptobranchus *and* Andrias*)*
Hula painted frog *(*Latonia nigriventer*)*
Purple frog *(Nasikabatrachus sahyadrensis)*
Jawless fish
Hagfish *(Myxinidae) family*
Bony fish
Arowana *and* arapaima *(Osteoglossidae)*
Bowfin *(Amia calva)*
Coelacanth *(the lobed-finned Latimeria menadoensis and Latimeria chalumnae)*
Gar *(Lepisosteidae)*
Queensland lungfish *(Neoceratodus fosteri)*
Sturgeons *and* paddlefish *(Acipenseriformes)*
Bichir *(Polypteridae) family*
Protanguilla palau

Mudskipper *(Oxudercinae)*
Sharks
Blind shark *(Brachaelurus waddi)*
Bullhead shark *(Heterodontus sp.)*
Elephant shark (Callorhinchus milii)
Frilled shark (Chlamydoselachus sp.)
Goblin shark *(Mitsukurina owstoni)*
Gulper shark *(Centrophorus sp.)*[239]

Given that essentially no evolutionary changes have happened with these creatures, then it is safe to assume that the idea that creatures change from one type of creature into another is absolutely *false*. Since we can show no evolution with these creatures, there are no aliens that have evolved either, we are not being visited by aliens from another planet, but again what we are seeing is in fact Demonic Aliens masquerading as aliens. It will be hard to accept that these creatures are not aliens, but once you examine them thoroughly, you will be able to see them for what they really are. When they try to say they are aliens, don't believe them! Even if they are aliens, they would not be here to be nice to us, but here to conquer us! Never in the history of the world has any society been "nice" to the other countries without expecting something from them. Even the United States exploits other countries for their oil or other natural resources. These Demonic Aliens want something from us, make no mistake about it! They, first of all, want to rule us, and then kill us and take over the world for their own. An example would be when the Conquistadors found the Aztecs and their gold, they didn't barter with them, but

---

[239] "Living Fossil," Wikipedia (last updated September 3, 2015), accessed 9/6/2015, https://en.wikipedia.org/wiki/Living_fossil.

killed as many of them as they could and just took their gold. It is the same with every culture and people, and now these creatures will do the same with us; they are just setting us up for now, until they can strike a decisive blow to destroy us! Why else would they be here? As soon as they can, they will make their move, deceive us into thinking they are here to help, and then conquer us and rule us! The great lie of evolution, rather than creation, is their main propaganda tool. People will fall prey to it, accept the implant, and then there is no hope for them. Even today we are being conditioned to take shots to combat Covid. We are now being told we need another to be safe. People will blindly take them and then, in the end, take the actual implant that will condemn them to Hell. While these shots are not the Mark of the Beast yet, they are the beginning of conditioning to follow the government and do whatever they say. The government is evil, and luckily the Founding Fathers wrote into the Constitution the right to bear arms, to protect ourselves from the tyrannical government that is coming.

Chapter 29
CONCLUSION

Whenever scientists start to discuss how we came into being they always say something like "Fourteen billion years ago…" but they are totally wrong in this first statement! Time does not exist if God did not make it! Nothing could exist if there is no God! You don't get to have anything if God didn't make it. Time, space, and matter are not in existence. To paraphrase Dr. Kent Hovand, all three of these things must come into existence at the same time, and what creates them has to be outside of them. If you have matter but no space, where would you put it? If you have space but no time, when would you put it? All three of these things must come into existence simultaneously or none of them could exist! What causes them to come into existence must also be outside of them, or it could not create them. Thus, without God, you can't have anything! God is spaceless, timeless, and immaterial! So, when they start with the concept that so many billions of years ago… they are wrong! You can't have time without it being created! You can't have matter without it being created, and you can't have space without it being created. **All of these things must happen at the same time. You have absolutely nothing! End of story!**

<table>
<tr>
<td>WITH GOD YOU GET:
SPACE, TIME, MATTER AND ALL OF CREATION</td>
<td>WITH EVOLUTION YOU GET:
NOTHING
0+0 = 0 (SPACE, TIME, AND MATTER DO NOT EXIST)
YOU DON'T HAVE BILLIONS OF YEARS FOR EVOLUTION TO TAKE PLACE BECAUSE YOU DON'T HAVE TIME, SPACE, OR MATTER!</td>
</tr>
</table>

Now scientists will argue that they have somehow had these items (Space, Time, and Matter) by default. But that is a lie. The Book of Genesis in the very first verse describes how we have them, God created them! I also need to credit Dr. Kent Hovand for pointing this out.

In the beginning **(Time)** God created the heaven **(Space)** and the earth. **(Matter)**
Genesis 1:1

In the beginning God created the heaven and the earth.
2 And the earth was without form, and void; and darkness was upon the face of the deep. And the Spirit of God moved upon the face of the waters.
3 And God said, Let there be light: and there was light.
4 And God saw the light, that it was good: and God divided the light from the darkness.
5 And God called the light Day, and the darkness he called Night. And the evening and the morning were the first day.
6 And God said, Let there be a firmament in the midst of the waters, and let it divide the waters from the waters.
7 And God made the firmament, and divided the waters which were under the firmament from the waters which were above the firmament: and it was so.
8 And God called the firmament Heaven. And the evening and the morning were the second day.
9 And God said, Let the waters under the heaven be gathered together unto one place, and let the dry land appear: and it was so.
10 And God called the dry land Earth; and the gathering together of the waters called he Seas: and God saw that it was good.
11 And God said, Let the earth bring forth grass, the herb yielding seed, and the fruit tree yielding fruit after his kind, whose seed is in itself, upon the earth: and it was so.
12 And the earth brought forth grass, and herb yielding seed after his kind, and the tree yielding fruit, whose seed was in itself, after his kind: and God saw that it was good.
13 And the evening and the morning were the third day.
14 And God said, Let there be lights in the firmament of the heaven to divide the day from the night; and let them be for signs, and for seasons, and for days, and years:

¹⁵ And let them be for lights in the firmament of the heaven to give light upon the earth: and it was so.

¹⁶ And God made two great lights; the greater light to rule the day, and the lesser light to rule the night: he made the stars also.

¹⁷ And God set them in the firmament of the heaven to give light upon the earth,

¹⁸ And to rule over the day and over the night, and to divide the light from the darkness: and God saw that it was good.

¹⁹ And the evening and the morning were the fourth day.

²⁰ And God said, Let the waters bring forth abundantly the moving creature that hath life, and fowl that may fly above the earth in the open firmament of heaven.

²¹ And God created great whales, and every living creature that moveth, which the waters brought forth abundantly, after their kind, and every winged fowl after his kind: and God saw that it was good.

²² And God blessed them, saying, Be fruitful, and multiply, and fill the waters in the seas, and let fowl multiply in the earth.

²³ And the evening and the morning were the fifth day.

²⁴ And God said, Let the earth bring forth the living creature after his kind, cattle, and creeping thing, and beast of the earth after his kind: and it was so.

²⁵ And God made the beast of the earth after his kind, and cattle after their kind, and every thing that creepeth upon the earth after his kind: and God saw that it was good.

²⁶ And God said, Let us make man in our image, after our likeness: and let them have dominion over the fish of the sea, and over the fowl of the air, and over the cattle, and over all the earth, and over every creeping thing that creepeth upon the earth.

²⁷ So God created man in his own image, in the image of God created he him; male and female created he them.

²⁸ And God blessed them, and God said unto them, Be fruitful, and multiply, and replenish the earth, and subdue it: and have dominion over the fish of the sea, and over the fowl of the air, and over every living thing that moveth upon the earth.

²⁹ And God said, Behold, I have given you every herb bearing seed, which is upon the face of all the earth, and every tree, in the which is the fruit of a tree yielding seed; to you it shall be for meat.

³⁰ And to every beast of the earth, and to every fowl of the air, and to every thing that creepeth upon the earth, wherein there is life, I have given every green herb for meat: and it was so.
³¹ And God saw every thing that he had made, and, behold, it was very good. And the evening and the morning were the sixth day.

Further, in the Book of Genesis where it talks about creation, all the plants, animals, stars, everything else is created in six days, but no aliens! Also, nothing is said about evolution taking place in order to get these things. We have been duped by Satan to usurp God and install him as God! Again, evolution apart from God is not possible as I have shown. But let's play the devil's advocate for a moment and say that aliens have evolved and are somehow visiting us from the planet K out in the far edge of our Universe. Why would they visit us? It could only be to exploit us and even kill us and take over our planet and rule over us. They are not going to be God's Messengers but would be Satan's Assassins! If they evolved, they don't follow God's Laws. It's those pesky 10 Commandments again that are not going to help us. They have no reason not to kill us because they obviously can! I am belaboring the point because it takes a while to sink in. If there are aliens, we have no hope. Luckily, there aren't!

How amazing this first verse really is! All three of the items necessary for creation are stated in it! Consider the Wisdom that drafted it! How could a person know this? Only God could make such a profound statement! So, when we talk about aliens somehow evolving outside of being created by God, which is not a valid argument, as God must have created them just like us, and God didn't tell us he created aliens. He made the stars, planets, and the Earth and everything in it, but if he made aliens, he would have told us. Since he didn't, they do not exist. He did make angels, however.

So, when we talk about aliens evolving, there is no mention of it in the Bible, and therefore, they were not created, nor did they evolve. However, we know he created angels, of which about 1/3 rebelled, and these are the ones that pilot the UFOs.

Now critics may say it doesn't talk about the creation of angels, and I can't find a verse that says that, but it talks about when the angels rebelled and were cast out:

> **⁷ And there was war in heaven: Michael and his angels fought against the dragon; and the dragon fought and his angels,**
> **⁸ And prevailed not; neither was their place found any more in heaven.**
> **⁹ And the great dragon was cast out, that old serpent, called the Devil, and Satan, which deceiveth the whole world: he was cast out into the earth, and his angels were cast out with him.** Revelation 12: 7 - 9

It does talk about God making man a little lower than the angels in Psalms 8 and Hebrews 2.

> 1.   Psalm 8:5
> **For thou hast made him a little lower than the angels, and hast crowned him with glory and honour.**
> 2.   Hebrews 2:7
> **Thou madest him a little lower than the angels; thou crownedst him with glory and honour, and didst set him over the works of thy hands:**
> 3.   Hebrews 2:9
> **But we see Jesus, who was made a little lower than the angels for the suffering of death, crowned with glory and honour; that he by the grace of God should taste death for every man.**

Obviously, God created the angels, and those that were cast out became the demons, and Satan is the god of this world. So again, the pilots of the UFOs are those demons cast out of Heaven to the Earth with bases at various places on the Earth. They may even fly them into outer space, but that doesn't make them aliens. A demon is still just a demon, and we are just a little lower than them. It talks about that here and in other places:

> **In whom the god of this world hath blinded the minds of them which believe not, lest the light of the glorious gospel of Christ, who is the image of God, should shine unto them.** 2 Corinthians 4:4

So, how will he blind us? By convincing us he is not a demon or Satan, but an alien- like on *Star Trek*, a "really good guy" and one of many alien races ready to "help" us with our problems. All we have to do is take their implant! We will be just like them, connected to the Internet with knowledge like they have. We will be God! We will no longer get sick, won't ever die, and don't have to obey those pesky 10 Commandments anymore! Lier! We will be condemned to Hell and will no longer be human!

From the Bible, how do I jump to the conclusion that UFOs have anything to do with it? One verse, in particular, drives me to that conclusion. It is this verse that talks about "strong delusion" and the context of the verse which states:

What else would make such an impact as aliens landing at the White House or being disclosed in a Presidential News Conference on all the television channels. That is a very strong delusion that would trash all the religions of the world, send the world into utter chaos. The Bible would be totally wrong, so forget about Jesus. It would set them up to be God and everyone would wonder after them. It is already happening when they said UFOs were real. Next, they will reveal the occupants of these crafts, and say they are aliens! Nothing else would even come close to this great deception. People would indeed follow them and do whatever they say, right down to taking the implant, and then Satan becomes the God he always wanted to be. We are doomed and those who follow after them are condemned when they take the implant.

I can't imagine it will happen any other way. Nothing else will be such a disruption to our systems. Aliens landing at the White House will shock the world. Then the Antichrist will be revealed. An alien-human hybrid that will have powers like the demons and will rule the world. He is most likely alive now, being groomed to rule, possibly at Area 51 or somewhere else. He will be revealed soon, I surmise after the Rapture. He will explain how it's the aliens that took the people out to their planet, not

Jesus. He will demonstrate his power by making "fire come down from Heaven."

[11] **And I beheld another beast coming up out of the earth; and he had two horns like a lamb, and he spake as a dragon.**

[12] **And he exerciseth all the power of the first beast before him, and causeth the earth and them which dwell therein to worship the first beast, whose deadly wound was healed.**

[13] **And he doeth great wonders, so that he maketh fire come down from heaven on the earth in the sight of men,**

[14] **And deceiveth them that dwell on the earth by the means of those miracles which he had power to do in the sight of the beast; saying to them that dwell on the earth, that they should make an image to the beast, which had the wound by a sword, and did live.**

[15] **And he had power to give life unto the image of the beast, that the image of the beast should both speak, and cause that as many as would not worship the image of the beast should be killed.**

[16] **And he causeth all, both small and great, rich and poor, free and bond, to receive a mark in their right hand, or in their foreheads:**

[17] **And that no man might buy or sell, save he that had the mark, or the name of the beast, or the number of his name.**

[18] **Here is wisdom. Let him that hath understanding count the number of the beast: for it is the number of a man; and his number is Six hundred threescore and six.**

Revelation 13: 11 - 18

I am writing over and over the same things in as many different ways that I can say it because it is too important to miss. All these things are going to happen soon and if you are here, you must not take the implant. Resist with everything you can. The "mark" is the implant, and it will condemn you to Hell. All your friends and family will tell you it is harmless, and for the first 3.5 years it will be, then it will change, and you will be powerless to resist what Satan commands you to do. He will be in charge of you completely, and you will be his slave.

Now, all this happens after the Rapture, we hope. But there is no guarantee. Do all Christians get raptured out, or are the bad (disobedient) Christians left behind, and only the good Christians raptured, or does all this happen with Christians here? In the past, they suffered for being Christians, were tortured, and killed just for being a Christian. It is happening even today in countries like China, Afghanistan, UAE, and Iran. Be a Christian in these places and you will be hunted down and killed! They are already hunting us down over there.

I wish I could say for sure that we will be taken out, but I can't! You don't know and no one else does either. Those Christians in these other countries aren't taken out before they are killed. In Rome, they were hung up on stakes or fed to lions! Today we have it too easy in the United States. It will get progressively worse, even maybe to being killed in our lifetimes. We have been shielded from persecution until very recently when non-believers started saying bad things about us. Soon it will not be a crime, no matter what they do to Christians. It happened to the Jews in WWII in Germany. It will happen here, but our only hope is it will wait until after the Rapture. But how could it be if there are no Christians to persecute? Looks like we have to be here during this time or else there is no one to put in FEMA camps that refuse to take the implant! It is going to happen very soon, and we may very well be here when it does. The Rapture might wait until halfway into the Tribulation it seems, but hopefully, it will be after I die, but if you are a young Christian be prepared! The time is very short now!

What if the aliens are actually demons and what if what I have written here is true, there are no aliens we are contending with, and demons are the only thing we need to worry about? Then it is just like it was from the beginning when Jesus came and died on the cross to save us. To say that is all wrong is going to destroy the fabric of our society, and other societies that believe in God. If there are aliens, then the Bible is wrong, or at least incomplete. Then Jesus didn't tell us what we need to know about life, and he must not be God because he left that fact out. There are no aliens! Jesus is God and the Bible is true, and we can trust what Jesus said. Evolution is the lie and the huge ages they ascribe for the Universe and time being responsible for our existence is not true, as I have already stated.

Look at what is happening now to show the Bible is true. At this time, we have the terrorist attacks that make men fearful like it says here:

> **Men's hearts failing them for fear, and for looking after those things which are coming on the earth: for the powers of heaven shall be shaken.** Luke 21:26

Also, the pestilence of Covid-19 is predicted in this verse:

> **[11] And great earthquakes shall be in divers places, and famines, and pestilences; and fearful sights and great signs shall there be from heaven.** Luke 21:11

We are clearly living in the last days now. These two predictions are coming true now, and there are many others that show the end is near. The Nation of Israel is a key to the end. As it states here:

> **[20] And when ye shall see Jerusalem compassed with armies, then know that the desolation thereof is nigh.** Luke 21:20

Never before has it been so perilous for Israel. All the other nations surrounding it are trying to snuff it out!

[14] Therefore, son of man, prophesy and say unto Gog, Thus saith the Lord GOD; In that day when my people of Israel dwelleth safely, shalt thou not know it?

[15] And thou shalt come from thy place out of the north parts, thou, and many people with thee, all of them riding upon horses, a great company, and a mighty army:

[16] And thou shalt come up against my people of Israel, as a cloud to cover the land; it shall be in the latter days, and I will bring thee against my land, that the heathen may know me, when I shall be sanctified in thee, O Gog, before their eyes.

[17] Thus saith the Lord GOD; Art thou he of whom I have spoken in old time by my servants the prophets of Israel, which prophesied in those days many years that I would bring thee against them?

[18] And it shall come to pass at the same time when Gog shall come against the land of Israel, saith the Lord GOD, that my fury shall come up in my face.

[19] For in my jealousy and in the fire of my wrath have I spoken, Surely in that day there shall be a great shaking in the land of Israel;

[20] So that the fishes of the sea, and the fowls of the heaven, and the beasts of the field, and all creeping things that creep upon the earth, and all the men that are upon the face of the earth, shall shake at my presence, and the mountains shall be thrown down, and the steep places shall fall, and every wall shall fall to the ground.

[21] And I will call for a sword against him throughout all my mountains, saith the Lord GOD: every man's sword shall be against his brother.

[22] And I will plead against him with pestilence and with blood; and I will rain upon him, and upon his bands, and upon the many people that are with him, an overflowing rain, and great hailstones, fire, and brimstone.

[23] Thus will I magnify myself, and sanctify myself; and I will be known in the eyes of many nations, and they shall know that I am the LORD. Ezekiel 38: 14 - 23

They want to destroy the fledgling nation that was just started again in 1948. They will have the capability to do it with nuclear weapons which is fast approaching. Only intervention by God himself will keep them from destroying it. That is what is going to happen when they try, Israel will defeat them!

It is most likely that this happens after the Rapture, but it is unclear when this will actually happen. Israel is always under threat. When Saddam Husain launched missiles against them in 1991, during the Gulf War, the United States was there to help protect them. Palestinian forces continue to fire rockets at them almost daily. This prophecy does not have to happen before the Rapture, but it could. The time is getting very short and coming to pass that was foretold in the Bible. This, among other reasons, leads me to believe there are no aliens we will have to contend with, but the same old demons we have always been contending with and these creatures that fly in these UFOs are demons and not aliens. I have no proof of any of this, but you should look again deeply into what they do to know they are not aliens. The number of pieces of evidence I suggest in this book should be enough to convict them of being demons, but if not, I still don't believe they are aliens. Even if they are aliens, they are still a threat to us, so demons or aliens they are still evil.

Nowhere in the Bible does it say anything about aliens attacking us, nor especially Revelation does not mention aliens when it talks about the last days and what is going to happen. It talks exclusively about demons and Satan and so we must determine that demons control the UFOs and not some benevolent aliens from some faraway planet. If there were aliens, would it not mention that fact somewhere? Everything scientists want us to swallow with evolution and aliens is conflicting with the Bible. Either Jesus is God, or he is not! Jesus is God and we are his special creation. When he did it is not critical to our creation, but abundance evidence suggests it was under 10,000 years ago. But no amount of time would account for our existence without creation, because again time would not exist.

I conclude, therefore, that aliens are mythical, like fairies, leprechauns, wolfmen, and mermaids. They could not have evolved over any amount of time and would have had to have been created by God, just like we were. The Universe being so large as it is, is designed specifically for the Earth, and any smaller the Earth could not exist with the life that it has. Demons, however, are very real and were angels that were created by God and then fell from their place to become the evil force that we contend with today. They have perfected lying to the point they make it very easy to believe them. When they are disclosed to us as aliens, billions of people will believe them that are not aware of the role of demons and how they deceive us. The world's governments will also go along with their deception because they will not acknowledge God or demons as being real, but aliens, they will believe exist. Anything found on Mars or another planet that looks alien could just as easily have been done by demons, so there is no issue there either. Even if they find "true aliens", or life on another planet, well that does not discount that these particular creatures are demons and not aliens. I say that with a huge amount of confidence because of what they do and how they act. Even if they are in fact aliens, they are demonic, at least controlled by Satan, as again he is the god of this world. The fact that they have bases here on the Earth where they operate out of shows they are not from outer space.

The idea that another dimension somehow exists where they could be operating out of is again too far "out there" to consider as a valid answer either. These demons will try to convince us of anything other than the truth, so be prepared to answer people that try to persuade you otherwise. I will now reiterate what I have said before to demonstrate the facts here with the reasons they are demons and not aliens:

True Aliens *are not* mentioned in the Bible, therefore do not likely exist.

No aliens are mentioned here, just a UFO-type craft, so this is not an alien craft. (Ezekiel chapter 10)

We could not have evolved, therefore, neither could aliens.

Nothing notates any aliens being "evolved" so this is still not notating aliens.

Demons *are* mentioned in the Bible; therefore, any discovered so-called *aliens* are demons.

The craft depicted in the Bible is angelic and not demonic, so it is not referring to aliens.

The lie they will state is that they "seeded" our planet, thus taking credit for God's creation.

This angelic craft is from God and not demonic.

The bases they have are here on Earth and not on some other planet, but they will lie and say they are from another planet. Government officials have stated there is no indication UFOs or UAPs are from another planet.

The craft depicted in the Bible is from God and not from another planet

They are abducting people, doing medical experiments on us, collecting genetic material, and all manner of nefarious acts, even collecting sexual materials (sperm and eggs) therefore they are not benevolent aliens but are shown to be demons.

While Elijah was "taken up" by God it does not say by a UFO. Enock was also "taken up" but it does not say by any type of craft.

This collecting of genetic material would be for the purpose of making hybrid-alien creatures which is a continuation of what was done before in Genesis when the demons "mated" with human females and produced giants.

Ezekiel was not captured and released with genetic material being taken but he saw the craft that was like a chariot.

Angels are a little higher than us, thus fallen angels (demons) would be more technologically advanced than we are, as these creatures are.

The craft Ezekiel saw was of a nature not described by what we see today, therefore, it is not an alien craft.

When they are disclosed, people will worship them and forget about God, like the Bible says, the whole world will wonder after the beast.

Ezekiel does not state to worship the craft and their occupants; therefore, we are to worship only God and not the craft or the occupants of the craft.

They are implanting people with devices, which falls into place with the Bible where everyone will be required to get the mark to buy or sell, which the implant could in fact be a type of "mark."

No implant is discussed in this chapter about when Ezekiel saw the craft, therefore this is not of aliens but of God.

Demonic entities control people and Demonic Alien Overlords flying above us would be in a position to do just that.

The creatures in the craft did not control him, but it was a vision of God and not of demons.

Other reasons demonstrating a young Universe.

No indication is given that these creatures in the craft are aliens, so it was of God and not of alien control.

They are sexually depraved.

The craft had no sexual context or depravity, but was displaying the Glory of God, therefore, is from God and angelic and not demonic.

They are not loving, therefore, are not of God!

The craft depicted in the Bible had an angelic loving nature, so this is of God and not aliens.

In the final conclusion here, there are no aliens we have found yet, and these UFOs are piloted by demons.

About the author

The life of James L. Kearns changed dramatically after he became a Christian on January 20, 1980. While James had a "normal" life up until that point, it was far from a happy, fulfilled life. He was withdrawn, drinking, and had no purpose. Even though as a young person he grew up at a private airfield his father owned and obtained a private pilot's license, built a Bradly GT kit car, and had motorcycles and fast cars, he was not happy. Since becoming a Christian he graduated from the University of Central Florida, with a BA in Communication, and has authored two other books, *Finding Proof of Jesus* and *The Chronicles of the Revelation* as well as numerous screenplays.